The Enjoyment of Math

The Enjoyment of Math

BY HANS RADEMACHER AND

OTTO TOEPLITZ

TRANSLATED BY HERBERT ZUCKERMAN

PRINCETON UNIVERSITY PRESS

PRINCETON, NEW JERSEY

Published 1957 by Princeton University Press,
Princeton, New Jersey
In the United Kingdom: Princeton University Press,
Chichester, West Sussex

Library of Congress Card No.: 57-627

ISBN 0-691-07958-7 (hardcover edition)
ISBN 0-691-02351-4 (paperback edition)

This is a translation from Von Zahlen und Figuren: Proben Mathematischen
Denkens für Liebhaber der Mathematik, by Hans Rademacher and Otto
Toeplitz, second edition originally published by Julius Springer, Berlin,
1933. Chapters 15 and 28 by Herbert Zuckerman have been added to the
English language edition.

Princeton University Press books are printed on acid-free paper and meet
the guidelines for permanence and durability of the Committee on
Production Guidelines for Book Longevity of the Council on Library
Resources

First Princeton Paperback edition, 1966
Second paperback and eighth hardcover printing, 1970
Seventh printing, for the Princeton Science Library, 1994

Printed in the United States of America

15 14 13 12 11 10 9 8 7

Preface

Otto Toeplitz, co-author of this book, died in Jerusalem on February 19, 1940, after having left Germany in the Spring of 1939. Toeplitz began his academic career in Göttingen as a disciple of David Hilbert, was then professor in Kiel and later in Bonn. His scientific work is centered around the theory of integral equations and the theory of functions of infinitely many variables, fields to which he has made lasting contributions.

The plan for this book arose at frequent meetings which the authors had, while Toeplitz was in Kiel and I was at the University of Hamburg. Both of us had repeatedly lectured about the subject matter of this book to a wider public. The manuscript was rewritten several times under mutual criticism. Toeplitz's great interest in the history of mathematics is still visible in the present edition. I remember with warm feelings the summer days in 1929 at Bonn, when together we put the last touches to the German manuscript. I am sure that Toeplitz would have been pleased and proud of the present English edition, a project of which he had often thought.

I wish to thank the translator Professor Herbert S. Zuckerman for his painstaking and understanding work. Not only do I admire his apt translation of the German text, but I also think that in his presentation of the content he has brought many of its arguments closer to the English speaking reader. He has added two chapters (15 and 28) to the book, which faithfully reflect the spirit in which this book was written.

My thanks go to my friends Emil Grosswald, D. H. Lehmer, and Herbert Robbins for help and valuable advice, and also to the publisher for his sympathetic cooperation.

Philadelphia, 1956   HANS RADEMACHER

CONTENTS

The Enjoyment of Math

Introduction

Mathematics, because of its language and notation and its odd-looking special symbols, is closed off from the surrounding world as by a high wall. What goes on behind that wall is, for the most part, a secret to the layman. He thinks of dull uninspiring numbers, of a lifeless mechanism which functions according to laws of inescapable necessity. On the other hand, that wall very often limits the view of him who stays within. He is prone to measure all mathematical things with a special yardstick and he prides himself that nothing profane shall enter his realm.

Is it possible to breach this wall, to present mathematics in such a way that the spectator may enjoy it? Cannot the enjoyment of mathematics be extended beyond the small circle of those who are "mathematically gifted"? Indeed, only a few are mathematically gifted in the sense that they are endowed with the talent to discover new mathematical facts. But by the same token, only very few are musically gifted in that they are able to compose music. Nevertheless there are many who can understand and perhaps reproduce music, or who at least enjoy it. We believe that the number of people who can understand simple mathematical ideas is not relatively smaller than the number of those who are commonly called musical, and that their interest will be stimulated if only we can eliminate the aversion toward mathematics that so many have acquired from childhood experiences.

It is the aim of these pages to show that the aversion toward mathematics vanishes if only truly mathematical, essential ideas are presented. This book is intended to give samples of the diversified phenomena which comprise mathematics, of mathematics for its own sake, and of the *intrinsic* values which it possesses.

The attempt to present mathematics to nonmathematicians has often been made, but this has usually been done by emphasizing the usefulness of mathematics in other fields of human endeavor in an effort to secure the comprehension and interest of the reader. Frequently the advantages which it offers in technological and other applications have been described and these advantages have been illustrated by numerous examples. On the other hand, many books have been written on mathematical games and pastimes. Although these books contain much interesting material, they give at best a very distorted picture of what mathematics really is. Finally, other books have discussed the foundations of mathematics with regard to their general philosophical validity. A reader of the following pages who is primarily interested in the pure, the absolute mathematics will naturally direct his attention toward just such an epistemological evaluation of mathematics. But this seems to us to be attaching an extraneous value to mathematics, to be judging its value according to measures outside itself.

In the following pages we will not be able to demonstrate the effects of the ideas to be presented on the domain of mathematics itself. We cannot consider the interior applications, so to speak, of mathematics, the use of the ideas and results of one field in other fields of mathematics. This means that we must omit something that is quite essential in the nature of the mathematical edifice, the great and surprising cross-connections that permeate this edifice in all directions. This omission is quite involuntary on our part, for the greatest mathematical discoveries are those which have revealed just such far-reaching interrelations. In order to present these interrelations, however, we would need long and comprehensive preparations and would have to assume a thorough training on the part of the reader. This is not our intention here.

In other words, our presentation will emphasize not the *facts* as other sciences can disclose them to the outsider but the *types of phenomena*, the *method of proposing problems*, and the *method of solving problems*. Indeed, in order to understand the great mathematical events, the comprehensive theories, long schooling and persistent application would be required. But this is also true with music. On going to a concert for the first time one is not able to appreciate fully Bach's "The Art of Fugue," nor can one immediately visualize the structure of a symphony. But besides the great works of music there are the smaller pieces which have something of true sublimity and whose spirit reveals itself to everyone. We plan to select such "smaller pieces" from the huge realm of mathematics: a sequence of subjects each one complete in itself, none requiring more than an hour to read and understand. The subjects are independent, so that one need not remember what has gone before when reading any chapter. Also, the reader is not required to remember what he may have been compelled to learn in his younger years. No use is made of logarithms or trigonometry. No mention is made of

6

differential or integral calculus. The theorems of geometrical congruence and the multiplication of algebraic sums will gradually be brought back to the memory of the reader; that will be all.

In the case of a small work of music it may not only be the line of melody with which it opens that makes it beautiful. A little variation of the theme, a surprising modulation may well be the climax of the whole. Only he who has listened attentively to the basic theme will fully perceive and understand this climax. In a similar sense our reader will have to "listen" readily and attentively to the basic motive of a problem, to its development, to the first few examples which illustrate each theme, before the decisive modulation to the cardinal thought takes place. He will have to follow the reasoning with a little more active attention than is usually required in reading. If he does this, he will find no difficulty in grasping the essential ideas of each subject. He will then get a glimpse of what a few great thinkers have created when they have occasionally left the realm of their comprehensive theoretical production and have built, from simple beginnings, a small self-contained piece of art, a fragment of the prototype of mathematics.

1. The Sequence of Prime Numbers

6 is equal to 2 times 3, but 7 cannot be similarly written as a product of factors. Therefore 7 is called a *prime* or *prime number*. A prime is a positive whole number which cannot be written as the product of two smaller factors. 5 and 3 are primes but 4 and 12 are not since we have $4 = 2 \cdot 2$ and $12 = 3 \cdot 4$. Numbers which can be factored like 4 and 12 are called *composite*. The number 1 is not composite but, because it behaves so differently from other numbers, it is not usually considered a prime either; consequently 2 is the first prime, and the first few primes are:

$$2, \ 3, \ 5, \ 7, \ 11, \ 13, \ 17, \ 19, \ 23, \ 29, \ 31, \ 37, \cdots.$$

A glance reveals that this sequence does not follow any simple law and, in fact, the structure of the sequence of primes turns out to be extremely complicated.

A number can be factored a step at a time until it is reduced to a product of primes. Thus $6 = 2 \cdot 3$ is immediately expressed as a product of two primes, while $30 = 5 \cdot 6$ and $6 = 2 \cdot 3$ gives $30 = 2 \cdot 3 \cdot 5$, a product of three primes. Similarly, 24 has four prime factors $(24 = 3 \cdot 8 = 3 \cdot 2 \cdot 4 = 3 \cdot 2 \cdot 2 \cdot 2)$, of which three happen to be the same prime, 2. In the case of a prime such as 5 one can only write $5 = 5$, a product of a single prime. By means of this step-by-step factoring, any positive whole number except 1 can be written as a product of primes. Because of this, the prime numbers can be thought of as the building blocks of the sequence of all positive whole numbers.

In the ninth book of Euclid's *Elements* the question of *whether the sequence of primes eventually ends* is raised and answered. It is shown that the sequence has no end, that after each prime yet another and larger prime can be found.

Euclid's proof is very ingenious yet quite simple. The numbers 3, 6, 9, 12, 15, 18, $\cdots$ are all multiples of 3. No other numbers can be divided evenly by 3. The next larger numbers 4, 7, 10, 13, 16, 19, $\cdots$, which are multiples of 3 increased by 1, are certainly not divisible by 3; for example, $19 = 6 \cdot 3 + 1$, $22 = 7 \cdot 3 + 1$, etc. In the same way the multiples of 5 increased by 1 are not divisible by 5 ($21 = 4 \cdot 5 + 1$, etc.). The same thing is true for 7, for 11, and so on.

Now Euclid writes down the numbers

$$2 \cdot 3 + 1 = 7$$
$$2 \cdot 3 \cdot 5 + 1 = 31$$
$$2 \cdot 3 \cdot 5 \cdot 7 + 1 = 211$$
$$2 \cdot 3 \cdot 5 \cdot 7 \cdot 11 + 1 = 2311$$
$$2 \cdot 3 \cdot 5 \cdot 7 \cdot 11 \cdot 13 + 1 = 30,031, \text{ etc.}$$

The first two primes, the first three primes, and so forth, are multiplied together and each product is increased by 1. None of these numbers is divisible by any of the primes used to form it. Since 31 is a multiple of 2 increased by 1, it is not divisible by 2. It is a multiple of 3 increased by 1 and hence is not divisible by 3. It is a multiple of 5 increased by 1, hence not divisible by 5. 31 happens to be a prime and it is certainly larger than 5. 211 and 2311 are also new primes, but 30031 is not a prime. However, 30031 is not divisible by 2, 3, 5, 7, 11, or 13, and hence its prime factors are greater than 13. As a matter of fact, a little figuring shows that $30031 = 59 \cdot 509$, and these prime factors are greater than 13.

The same argument may be applied as far as one wants to go. Let p be *any* prime and form the product of all primes from 2 to p; increase this product by 1 and write

$$2 \cdot 3 \cdot 5 \cdot 7 \cdot 11 \cdots p + 1 = N.$$

None of the primes 2, 3, 5, $\cdots p$ divides N, so either N is a prime (certainly much greater than p) or all the prime factors of N are different from 2, 3, 5, $\cdots p$, and hence greater than p. In either case, a new prime greater than p has been found. No matter how large p is there is always another larger prime.

This part of Euclid is quite remarkable, and it would be hard to name its most admirable feature. The problem itself is only of theoretical interest. It can be proposed, for its own sake, only by a person who has a certain inner feeling for mathematical thought. This feeling for mathematics and appreciation of the beauty of mathematics was very evident in the ancient Greeks, and they have handed it down to later civilizations. Also, this problem is one that most people would completely overlook. Even when it is brought to our attention it appears to be trivial and superfluous, and its real difficulties are not immediately apparent. Finally, we must admire the ingenious and simple way in which Euclid proves the theorem. The most natural way to try to prove the theorem is not Euclid's. It would be more natural to try to find the next prime number following any given prime. This has been attempted but has always ended in failure because of the extreme irregularity of the formation of the primes.

Euclid's proof circumvents the lack of a law of formation for the sequence of primes by looking for *some* prime beyond instead of for the *next* prime after p. For example, his proof gives 2311, not 13, as a prime past 11, and it gives 59 as one past 13. Frequently there are a great many primes between the one considered and the one given by the proof. This is not a sign of the weakness of the proof, but rather it is evidence of the ingenuity of the Greeks in that they did not try to do more than was required.

As an illustration of the complexity of the sequence of primes, we shall show that there are large gaps in the series. We shall show, for example, that we can find 1000 consecutive numbers, all of which are composite. The method is closely related to that of Euclid.

We saw that $2 \cdot 3 \cdot 5 + 1 = 31$ is not divisible by 2, 3, or 5. If two numbers are divisible by 2 then their sum is certainly divisible by 2 as well. The same is true for 3, 5, etc. Now, $2 \cdot 3 \cdot 5$ is divisible by 2, 3, and 5, so $2 \cdot 3 \cdot 5 + 2 = 32$ is divisible by 2, $2 \cdot 3 \cdot 5 + 3 = 33$ by 3, $2 \cdot 3 \cdot 5 + 4 = 34$ by 2, $2 \cdot 3 \cdot 5 + 5 = 35$ by 5, and $2 \cdot 3 \cdot 5 + 6 = 36$ by 2. Therefore none of the numbers 32, 33, 34, 35, 36 is a prime. This argument fails for the first time for $2 \cdot 3 \cdot 5 + 7 = 37$, which is not divisible by 2, 3, or 5.

In the same way we can find 1000 consecutive composite numbers. Let p be the first four digit prime (1009) and form the 1000 numbers

$$2 \cdot 3 \cdots p + 2, \qquad 2 \cdot 3 \cdots p + 3 \cdots, \qquad 2 \cdot 3 \cdots p + 1001.$$

Each of the numbers 2, 3, 4, 5, $\cdots$, 1001, is divisible by one of the primes 2, 3, $\cdots$, p and so is $2 \cdot 3 \cdots p$. Therefore each of the numbers listed is also divisible by one of 2, 3, $\cdots p$, and hence is not a prime itself. We have thus found 1000 consecutive numbers none of which is prime, and consequently there is a gap of at least 1000 numbers in the sequence of primes.

Naturally a gap of such length will not occur until we have gone quite far along in the sequence of primes, but if we go far enough we can, by the same method, find gaps as long as we may desire.

This problem and Euclid's are very similar both in nature and proof, yet the question of gaps in the sequence of primes was not considered by the Greeks. It was taken up by modern mathematicians along with a great number of other related problems. Most of these other problems are not as easy to solve; some remain unsolved at the present time, while others have led to entirely new fields of mathematics.

Let us consider one of these further problems which we can still handle by the same methods and which will give a certain insight into some of the others. The multiples of 3 are 3, 6, 9, $\cdots$, and these numbers increased by 1 viz. 4, 7, 10, $\cdots$ have occurred above. The remaining numbers 2, 5, 8, 11, 14, 17, 20, 23, $\cdots$ are the numbers which have the remainder 2 when divided by 3. Do these last numbers contain infinitely many primes? That is, does the sequence 2, 5, 11, 17, 23, $\cdots$ never end? We shall prove that there is an infinite number of these primes.

First we must show that if any two of the numbers 1, 4, 7, 10, 13, $\cdots$ are multiplied together, the product is again one of these same numbers. Each of the numbers is a multiple of 3 increased by 1, say $3x + 1$. If a second number is $3y + 1$, then their product is

$$
\begin{aligned}
(3x + 1)(3y + 1) &= 9xy + 3y + 3x + 1 \\
&= 3(3xy + y + x) + 1,
\end{aligned}
$$

(1)

which is again a multiple of 3 increased by 1 and hence is in our original set of numbers.

Now if any of the numbers 2, 5, 8, 11, $\cdots$ is split into its prime factors, at least one of the prime factors must again be one of 2, 5, 8, 11, $\cdots$. For example, $14 = 2 \cdot 7$ has the factor 2 and $35 = 5 \cdot 7$ has 5. To show that this is always true, we note that each of the prime factors must be either a multiple of 3, in the sequence 1, 4, 7, 10, $\cdots$, or in the sequence 2, 5, 8, $\cdots$. The only multiple of 3 that is a prime is 3 itself, and it does not divide our chosen number. If all the prime factors were of the kind 1, 4, 7, 10, $\cdots$, then by the above remark (1) our number would again be of this kind. Since it is not of this kind, at least one of its prime factors must be one of the kind 2, 5, 8, 11, $\cdots$.

We can now proceed as in Euclid's proof except that we consider

$$2 \cdot 3 \cdot 5 \cdot 7 \cdot 11 \cdots p - 1 = M$$

in place of

$$2 \cdot 3 \cdot 5 \cdot 7 \cdot 11 \cdots p + 1 = N.$$

M is a multiple of 3 *decreased* by 1, which means that it is one of 2, 5, 8, 11, $\cdots$. Just as with N, it is clear that M is not divisible by any of the primes 2, 3, 5, 7, 11, $\cdots$, p. Either M is a prime (greater than p), or all its prime factors are greater than p. In the latter case at least one of the prime factors is one of 2, 5, 8, $\cdots$, and hence in both cases we have found a prime of this kind that is greater than p. Therefore the sequence 2, 5, 8, $\cdots$ contains an infinite number of primes.

This leaves open the question whether the sequence 1, 4, 7, 10, 13, $\cdots$ contains an infinite number of primes. It is quite possible that 2, 5, 8, . . . could contain infinitely many primes while 1, 4, 7, $\cdots$ contained only a finite number. The fact is that the latter sequence also contains infinitely many primes, but the proof of this requires completely different methods. In a later chapter we shall gain some insight into these methods. The reason for mentioning these last problems was to point out how the problems and methods of modern mathematics are related to those of the Greeks. This is not only true in isolated cases, but in whole areas of modern mathematics, areas of very great interest in which research is still being carried on.

2. Traversing Nets of Curves

A streetcar company decides to reorganize its system of routes without changing the existing tracks. It wishes to do this in such a way that each section of track will be used by just one route. Passengers will be allowed to transfer from route to route until they finally reach their destinations. The problem is: *how many routes must the company operate in order to serve all sections without having more than one route on any section?*

For a very small city with car lines as in Fig. 1, the problem is quite simple. One route could go from A to B and a second from C to D, both passing through K. Or a line could go from A through K to D and a second from B through K to C. Finally, a line could go from A through K to C and another from B through K to D. Each of these possibilities necessitates two routes. Of course the company could set up a new transfer point at R, end a route there, and run a shuttle from R to B. But this would only increase the total number of routes needed. By adding more transfer points the number of routes could be increased indefinitely, but we are not interested in doing this, since the original problem asks for the *smallest* possible number of routes.

Fig. 2 is still a fairly simple network of lines, but the problem is more complicated in this case. One possibility is for one route to run from *A* around through *B, C, D, E,* and back to *A* (a closed route).

Fig. 1

A second route might go from *A* through *F, G, H,* to *D.* Three more routes, *BF, EG, CH,* would be needed, making five in all. However, this is not the best possible arrangement. The first two routes could be combined into a single one running from *A* around through *B, C, D, E,* and back to *A,* and then on through *F, G,* and *H* to *D.* This cuts the number of routes to four, but *would it be possible to set up just three routes?*

In the network of Fig. 3 one route might proceed from *A* through *B, C, D,* and *E,* back to *A,* and then on through *F* to *B.* That would leave three sections, *CF, DF, EF,* of which two could be

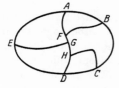

Fig. 2

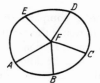

Fig. 3

combined into a single route *CFD,* leaving *EF* as a route by itself. In this case the first two routes pass through *F* while the third ends at *F.* The question still remains: *would two routes do just as well?*

It would not be unduly difficult to list all the possible routes for Figs. 2 and 3 (as we did for Fig. 1) in order to be sure we have the minimum number of routes. However, this would be quite a long procedure for a fairly complicated network, and in any case it would not be a very interesting problem. Also, it would no more be mathematics merely to list all the possibilities for any given network than it is mathematics to multiply together two seven-digit numbers. The spirit of mathematics dictates that we seek out the essentials of the problem and use them to solve it.

The essential thing in this problem is quite simple. We must consider where the *ends* of the various routes must lie. There must be an end wherever a section has a free end as at A, B, C, and D in Fig. 1. Since in this case there must be four ends, and since each route has no more than two ends (a closed route has no ends of course), it is clear that there must be at least two routes between these four ends. By means of a single consideration we have obtained the same result as was obtained above by considering all the possible cases.

In Fig. 2 there are no free ends, but here there are places such as A where three sections come together. At such a place at least one route must end, since two routes can never be on the same section. In Fig. 2 there are eight places of this kind, so there must be at least eight ends of routes. Eight is an even number, the number of ends of four routes. Therefore there must be at least four routes; and, as we have already seen, four routes will do.

In Fig. 3 there are five junctions of the kind we have been discussing, and a sixth point F where not three but five sections come together. We shall call a point like F a junction of the fifth order. At F two pairs of sections might be joined to form routes, leaving one over, as was done in the first discussion of Fig. 3. The same thing would be true for any junction of *odd* order; one section is always left over, and hence there must be at least one end there. We now see that Fig. 3 requires at least six ends (an even number) and hence at least three routes.

For any network, no matter how complicated, we can easily count the junctions of odd order and divide by 2 to obtain the least possible number of routes. In the three examples so far, the number of junctions of odd order was always even, and it turned out that half this number of routes was not only necessary but also sufficient to meet the requirements of the problem. We would now like to determine whether or not the number of junctions of odd order is always even, and whether half this number of routes will always do for any given network.

No matter how complicated a network is, it is certainly possible to find some system of routes having no more than one route on any section. All one needs to do, in fact is to establish separate shuttle lines on each section between each pair of junctions. However, this will certainly require far too many routes. There will be other systems using fewer routes, and we are looking for the best system, the minimum number of routes needed. Among all the conceivable systems there will be certain ones that are best — such that no other system requires fewer routes.

It is clear that a system of routes is certainly not best if it includes extra ends like the transfer point R of Fig. 1. Neither is it best if more than one route ends at a point like F in Fig. 3, for if one route came through C to F and another came through D to F, then both could be connected at F to form a single route, and this would decrease the total number. In order for a system to be best, the sections at each junction must wherever possible be paired into routes. This means that just one route will end at a junction of odd order, none at a junction of even order, and the total number of ends will be the number of junctions of odd order in the network.

We must still decide whether a system which is best can contain a closed route. Such a closed route was mentioned in our discussion of Fig. 2, but we connected it at A to the route $AFGHD$ and decreased the number of routes from five to four. The resulting route $ABCDAFGHD$ is not closed--it has the two ends A and D. This same reduction can be made whenever a closed route contains a junction of odd order, and a similar reduction can be made if the closed route contains only junctions of even order. Let A be such a junction (Fig. 4) on a closed route (shown here in the form of a

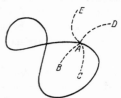

Fig. 4

figure eight). Some other routes through A are shown as dotted curves and they may continue on in any way. If the system is best no route can end at A, so the line from B through A continues on, let us say, through E. We can then combine this line and the closed route into a single route running through B to A, around the closed route back to A, and then on through E. Since this reduces the number of routes, the original system could not have been best. However, we should note that in this reduction the newly formed route may again be closed, since the original line through BAE may have continued on and back to B. If this is the case, further reductions may be possible. If at some stage the new closed route contains a junction of odd order, then the next reduction will yield a route that is not closed. Otherwise all the junctions of the original network must have been of even order and the system will finally be reduced to a single closed route.

To sum up our results, we have seen that if a system is best, routes end only at junctions of odd order and only one route ends at each such point. Also, there will be a closed route in the system only if all the junctions of the original network are of even order, and then the closed route will traverse the entire network. In a best system, the number of junctions of odd order is equal to the number of ends of routes, and is therefore an even number. Furthermore, the minimum number of routes is half the number of junctions of odd order, except in the case where all junctions are of even order, when the minimum is one (closed) route.

3. Some Maximum Problems

1. Let us compare the areas of various rectangles of two-inch perimeter. Some are shown in Fig. 5. The width of each rectangle must be less than 1 inch, and the closer it is to 1 inch the smaller is the height and also the area. If the height is close to 1 inch, then the width and again the area are very small. The intermediate rectangles have larger area, and one might ask which of the rectangles has the largest area. This is a maximum problem. It is probably the simplest and oldest of all such problems, and so perhaps the most suitable one to use as an introduction.

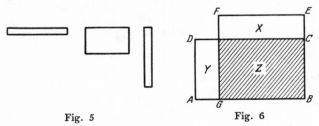

Fig. 5   Fig. 6

This problem is solved in Euclid, Book VI, theorem 27. Our proof will use the same principles and will differ from Euclid's only in its presentation. The rectangle $ABCD$ of Fig. 6 is supposed to represent any rectangle with a fixed perimeter P. The square

BEFG has each side of length $\frac{1}{4}P$, so its perimeter is also *P*. We assert that the square is the answer to our problem, that its area is greater than the area of any rectangle (not a square) *ABCD* having the same perimeter. In the figure the shaded rectangle $\mathcal{Z}$ is part of both the original rectangle and the square. The square also contains the area *X*, while the original rectangle is made up of $\mathcal{Z}$ and *Y*. Now $AB + BC$ is half the perimeter of the rectangle, and $GB + BE$ is half the perimeter of the square, so these two are equal; $AB + BC = GB + BE$. This can be written as $AG + GB + BC = GB + BC + CE$, from which we have $AG = CE$. Thus the rectangle *X* is just as high as *Y* is wide. However, the width of *X* is one side of the square, while the height of *Y* is *part* of the side of the square, and is therefore smaller. If two rectangles (Fig. 7) are the

Fig. 7

same length in one dimension, then the one with the larger other dimension has the larger area, i.e., *X* is larger than *Y*. Therefore $X + \mathcal{Z}$ is larger than $Y + \mathcal{Z}$, and the square has larger area than the rectangle. The height of *Y* would no longer be a part of the side of the square only if *ABCD* were a square itself, and then the square *BEFG* would itself be the original rectangle *ABCD*. Therefore *the square is larger than all other rectangles with the same perimeter.*

This result has been stated as the Greeks would have put it. It can also be written as an algebraic formula, which is the way we would put it in modern mathematics. Let *x* and *y* be the dimensions of the rectangle in inches. The area of the rectangle may then be given in square inches and its perimeter is $x + y + x + y = 2(x + y)$ inches. Therefore the square has sides $\frac{1}{2}(x + y)$ inches in length and an area of $\left(\dfrac{x + y}{2}\right)^2$ square inches. Thus, *if x and y are any two positive numbers*, the result becomes [1]

$$xy \leq \left(\frac{x + y}{2}\right)^2,$$

or

[1] The symbol $<$ means, and is read, "is less than"; $\leq$ means "is less than or equal to." Thus, $3 < 5, 3 \leq 5, 3 \leq 3$. Likewise, the symbol $>$ means "greater than," and $\geq$ means "greater than or equal to."

$$\sqrt{xy} \leqq \frac{x+y}{2}.$$

Verbally, this might be stated: *the geometric mean of two numbers is always less than their arithmetic mean.* The two are equal only if *x* and *y* are equal.

2. We have now seen what is meant by a maximum problem and its solution. First a solution is proposed, after which it is proved that the figure named in this solution exceeds all the other figures with which it is to be compared with regard to a given property (in this case, area). It is now possible to turn to the main problem of this section: *to find the triangle of largest area that is inscribed in a given circle.* It is likely that this problem was discussed, if not solved, at the time of Plato, a century before Euclid. However, neither Euclid nor more modern books give the following solution, which could easily have been understood and discovered by the Greeks.

Besides the original triangle ABC, we consider the equilateral triangle $A_0B_0C_0$ inscribed in the same circle or another equal circle (Fig. 8). The area of $A_0B_0C_0$ is completely determined, since the

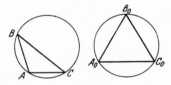

Fig. 8

equilateral triangle is definitely fixed except that it may be turned. We claim that the equilateral triangle is the solution to our problem, that its area is greater than the area of any other inscribed triangle.

We first note that the circumference of our circle is split into three equal arcs by the equilateral triangle, while the other triangle breaks it up into three unequal arcs. Of these last three arcs, one must be larger than one-third of the total circumference of the circle. Otherwise the three arcs would each have to be exactly one-third the circumference in order to make up the whole. The triangle ABC would then be equilateral, but we have assumed that this is not the case. In the same way, one of the arcs is smaller than one-third the circumference. The third arc may be larger or smaller than one-third the circumference; we are not able to decide which, but it will not affect our argument.

Let us suppose that the triangle has been labeled in such a way that the arc AB is less than one-third the circumference, and arc BC is more. As in Fig. 9, we measure an arc CB'' with a length equal to that of AB along CB. The triangle CAB'' is the mirror image of ACB reflected in the diameter of the circle perpendicular to AC. We also measure the arc AB', which is equal to one-third the circumference in the same direction from A as B. Since AB is less than one-third the circumference, B' will be past B. However, B' will not be past B'', that is, it will fall between B and B''. If it were past B'', the arc AB' would be larger than AB'', which is the reflection of CB. But CB was greater than one-third the circumference, while we made AB' exactly one-third the circumference.

Now since B' lies between B and B'', it is higher than B. The triangles ACB and ACB' have the same base AC, and the altitude

Fig. 9

Fig. 10

of ACB is less than the altitude of ACB'. Since the area of a triangle is one-half the altitude times the base, ACB has less area than ACB'. We have found a new inscribed triangle ACB' of greater area than the original triangle ABC and with one side AB' equal to the side of an inscribed equilateral triangle.

It may happen that ACB' is an equilateral triangle. This will be the case if the arc AC cut off by the original tirangle was exactly one-third the circumference. Then the proof that the equilateral triangle is larger than the original triangle would be complete. If ACB' is not equilateral we consider AB' as the base of the figure just as AC was before. To do this we turn the figure around as in Fig. 10 until AB' is at the bottom. We can now treat $AB'C$ with base AB' exactly as we did ACB with base AC. We will end up with the triangle $AB'C'$, which is equilateral since both $B'C'$ and AB' are each one-third the circumference of the circle. Because $A_0B_0C_0$ of Fig. 8 and $AB'C'$ are equilateral triangles inscribed in equal circles, their areas are equal. We have shown that the area of $AB'C'$ is larger than that of $AB'C$, which in turn is larger than the original triangle ABC, always assuming that ABC is not equilateral.

20

This completes the proof; the equilateral triangle has a larger area than any other triangle inscribed in the same circle.

3. The previous result is a particular case of a more general statement that can be proved in a similar way. We shall show that *of all polygons of n sides inscribed in a given circle, the regular polygon is largest.* If $n = 3$ the polygon is a triangle and this is our previous result.

In order to prove this we need to notice just one more simple fact. If we are given any polygon inscribed in a circle (Fig. 11), we can inscribe another polygon in the circle having the same sides but in any other order. All we need to do is draw the radii to the corners of the polygon and cut the circle into sectors along these radii.

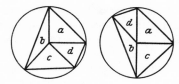

Fig. 11

These sectors can then be rearranged in any desired order. It is obvious that the new polygon has the same area as the original one.

The proof now goes just about as before. In the first place, if the polygon is not regular, there must be a side that cuts off an arc less than one n-th of the circumference, and one that cuts off an arc larger than one n-th. In the case of a triangle any two sides are always next to each other, but if n is greater than three the two sides we are interested in may not be next to each other. However, we have just seen that we can draw a new polygon in the same circle, with the same sides and area, but with the two particular sides next to each other. Suppose the smaller side is called AB, the larger BC. We can measure off one n-th of the circumference from A in the direction of B and call its end B'. As before, B' lies between B and the mirror image B'' of B. Then the new polygon with B' in place of B and all the other corners the same has a larger area than the original polygon. Also, one side, AB', of the new polygon is the length of a side of a regular polygon. We can apply this same procedure to the $n - 1$ remaining sides, always rearranging the order of the sides if necessary. Continuing a step at a time, we will finally arrive at a regular polygon, since a new side becomes the correct length at each step. Because the area is increased, each time, the regular polygon has a larger area than the original polygon.

In a similar way it can be shown that of all polygons of n sides *circumscribed* about a circle, the regular polygon has *least* area.

4. Incommensurable Segments and Irrational Numbers

The measurement of length, area, and volume is at the root of all geometry. To measure one line segment by another, we see how many times one goes into the other. This is simple enough if it goes in exactly without leaving a remainder. If the smaller segment does not go into the larger one exactly, then we look at the remainder. It may happen that the remainder is one-half, one-third, two-thirds, or some other similar fractional part of the segment we are using as a measure. If so, we have a sort of substitute measure, a fractional part of the segment that goes exactly into both the segment being measured and the one that is used as a measure. This new segment is a "common measure" for the two original segments.

The earliest geometrical problems undoubtedly involved a common measure. For example, if a rectangle has sides 3 and 4 inches in length, then by the Pythagorean theorem the square constructed on the diagonal has area

$$3^2 + 4^2 = 9 + 16 = 25 \text{ sq. in.}$$

and the diagonal is consequently 5 inches long. Since the smaller side of the rectangle and the diagonal have a 1 inch segment as a common measure, they are in the proportion 3 : 5.

It is quite natural to try to do the same thing for a square and to look again for a common measure for the side and diagonal. If we attempt to do this, we find a remainder (Fig. 12) and are forced to try smaller and smaller fractional parts of the segment. The

Fig. 12

question arises as to whether one can eventually find a fine enough measure or whether there just isn't any such measure, that is, whether the two segments are *"commensurable"* or *"incommensurable"*. This leads to the question whether a line segment can be divided into arbitrarily fine divisions or whether there is a limit to the possible divisions. Is the line made up of a very large number of small indivisible parts? Does it have an "atomic structure"? The conception of the atomic structure of matter is attributed to Democritus, who lived just before Plato. However, there is a difference between this and the atomic structure of the line. One can easily consider a line as "continuous," i.e. capable of arbitrarily fine division, and still suppose that physical material is strung along it in a series of atoms. A fragment attributed to Anaxagoras has also been preserved from about the same time; this asserts, in effect, that the line is continuous. The fact that only this fragment has been handed down does not signify that it was just an accidental utterance. Rather, it was a controversial thesis that led to much discussion, and it is representative of a time when mankind was making real steps toward the solution of this basic problem.

We can imagine the tremendous effect of the more far-reaching discovery that *the side and diagonal of a square are incommensurable.* This discovery is attributed to the Pythagoreans, a secret society of Southern Italy about whom very little is definitely known. According to a legend, the Pythagorean who made these investigations public atoned by perishing in a shipwreck. Perhaps this is more allegory than truth, since it may refer to the shattering effect that the discovery of irrationals had on the foundations of contemporary thought. In any event we have Plato's own report in his *Laws* of how this discovery excited him when he first learned of it.

We shall give two proofs of this fact without considering the interesting historical question of which proof is the older. The second was given not only by Euclid but even by Aristotle. The first, apparently older, is of the type usual with the Greeks and, in spirit, belongs to the tenth book of Euclid.

For the first proof we must begin by noticing certain facts of an elementary geometric nature. The whole argument is easily recognized as arising from a vain attempt to find a common measure by measuring the side off along the diagonal and then continuing to attempt to find a suitable fractional part. We measure off the side along the diagonal from B (Fig. 13). It goes in only once and we can call the point where it ends D. The line $B'D$ has been drawn perpendicular to BD and B' is the point where it crosses AC. Also B and B' have been connected. We have $BA = BD$, $BB' = BB'$, angle $BAB' = $ angle BDB', the last because each is a right angle. There-

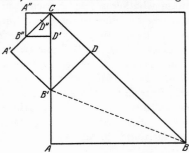

Fig. 13

fore by one of the congruence theorems, the triangles BAB' and BDB' are congruent and consequently AB' and DB', being corresponding sides, are equal. Also angle $B'CB$, between a side of the square and a diagonal, is half a right angle, and since angle CDB' was constructed a right angle, just half a right angle is left over for the third angle of the triangle CDB'. Therefore CDB' is an isosceles right triangle and its legs DB' and DC are equal. Combining what we have proved, we have

$$(1) \qquad AB' = B'D = DC.$$

We now erect a perpendicular $A'C$ to the diagonal at C and make it equal to DB'. When A' is connected to B', $A'B'CD$ forms a new square smaller than the original one. The whole process can now be applied to the new square, its side is measured off along its diagonal $B'C$ from B to give D', and the line $B''D'$ drawn perpendicular to $B'C$. As before, we have

$$(2) \qquad A'B'' = B''D' = D'C.$$

Clearly this can be repeated again and again, and each time there will be a remainder on the diagonal which can be used as a side of the next square. Although the process will never end, the remainders (which are never zero) become smaller at each step:

$$(3) \qquad CD > CD' > CD'' > CD''' > \cdots.$$

Each of these remainders is the difference between the diagonal and side of the corresponding square:

$$(4) \quad CD = CB - AB, \quad CD' = CB' - A'B', \quad CD'' = CB'' - A''B'', \cdots.$$

This completes the geometric preliminaries for the first proof. The proof itself will be an indirect one. We assume that the side and diagonal are commensurable and show that this leads to an impossible situation. If the two are commensurable they have a common measure, a segment E that goes exactly into both the side and the diagonal. Now if any two segments are exact multiples of E, their difference is also an exact multiple of E. Therefore, if CB and AB are exact multiples of E, so is CD by (4). Also, by (1) we have $CB' = CA - B'A = AB - CD$, which is a difference of multiples of E and hence an exact multiple of E itself. The second square of Fig. 13 has side $A'B' = CD$ and diagonal CB', both of which are multiples of E. From the first square we have gone to the second and we can go on in the same way; from $A'B'$ and CB' we find that CD', $A''B''$, and CB'' are multiples of E, and so on through the further squares.

We now come to the contradiction. If CB and AB are exact multiples of E, then we have seen that CD', CD'', CD''', $\cdots$ are also exact multiples of E. But (3) shows that these multiples of E get smaller and smaller without stopping, though they are never zero. This is not possible since, for example, if CD were $1000\,E$, then CD', a smaller multiple of E, would be at most $999\,E$, etc., until at the latest the 1001st term would be less than E. It would then be zero, since it is a multiple of E less than E. This contradicts the fact that no term of (3) is zero.

The *second proof* is much simpler, and the arithmetical preparation for it is shorter than the geometric preliminaries of the first proof.

We first consider even and odd numbers. An even number is twice some other, so that it can be written $2x$. An odd number is an even number increased by 1, and can be written $2x + 1$. The square of an odd number is always odd, for

$$(2x + 1)^2 = 4x^2 + 4x + 1 = 2(2x^2 + 2x) + 1$$

is again twice a number increased by 1. From this we can immediately prove:

Lemma [2] 1. *If the square of a number is even then the number itself is even.* If the number were odd, as we have just seen, its square would also be odd. It is even easier to prove:

Lemma 2. The square of an even number is always divisible by 4. Thus $(2x)^2 = 4x^2$ is 4 times x^2, or simply $4g$.

[2] A lemma is a proposition not important enough to be called a theorem, but which is to be used in the later proof.

The main proof is again indirect. We again suppose that the side and diagonal of the square have a common measure E. Let the diagonal be d times E, the side s times E. Then, applying the Pythagorean theorem to one of the right triangles formed by two sides of the square and the diagonal, we have

(5) $$d^2 = s^2 + s^2, \quad d^2 = 2s^2.$$

We can suppose that d and s have no common factors since they can always be reduced by a new choice of E. For example, 10 and 16 have the common factor 2, but they can be reduced to 5 and 8 by doubling the size of the common measure E. From now on we shall assume that such a reduction has been made.

From (5) we see that d^2 is twice a number and hence is even. Lemma 1 then shows d is also even. Consequently s must be odd, since otherwise d and s would have the common factor 2 in contradiction to the fact that they are reduced. However, since d is even, lemma 2 shows that d^2 is divisible by 4, $d^2 = 4g$, and $2s^2 = 4g$ or $s^2 = 2g$. Therefore s^2 is even and so is s (by another application of lemma 1). This contradicts the fact that we have just shown (that s is odd), and hence shows that the original assumption, that s and d have a common measure, is false.

The essence of both proofs is that a decreasing sequence of whole positive numbers must finally come to an end. In the first proof it is apparent in our discussion of the sequence (3). In the second proof it is hidden but really included in the remarks concerning the reduction of the numbers d and s. The proof of the fact that the reduced form of two numbers can be found depends on a series of decreasing steps.

In the current mathematical notation, formula (5) would be written

$$\left(\frac{d}{s}\right)^2 = 2.$$

Likewise, our final result would be written: There is no fraction (no rational number) $x = \dfrac{d}{s}$ whose square is 2. This can also be stated as: There is no rational number which is equal to $\sqrt{2}$, or finally, $\sqrt{2}$ is an "irrational number".

5. A Minimum Property of the Pedal Triangle

We shall again consider a problem of the kind discussed in Chapter 3, but this time it might more properly be called a minimum problem. It will serve to introduce mathematical methods that are highly refined yet clear and simple. The theorem and the proof we shall give here are the work of H. A. Schwarz. Although the theorem is only a relatively minor mathematical problem, it shows how this great mathematician's genius manifests itself, equally in relatively trivial and extremely significant work.

1. Before considering our main theorem, let us look at a very simple problem concerned with the law of reflection of light. It is well known that if a light ray starts at A (Fig. 14) and strikes a mirror g, it is reflected toward B in such a way that the angle of

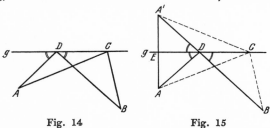

Fig. 14   Fig. 15

incidence and the angle of reflection are equal. What we want to prove is that the path ADB that the light ray chooses is the shortest of all possible paths from A to B that touch the mirror g. It is the path that a steamboat should take if it were required to go from a place A to another place B and to touch at the bank g on the way. We shall not go into the question of why a light ray, which does not have the power of reasoning, chooses the same path that would be chosen by the pilot of the steamboat after considerable thought. All we shall prove is the purely mathematical fact that *the path ADB with equal angles of incidence and reflection is shorter than any other path ACB.*

The proof depends on a device that seems artificial from a mathematical standpoint, but which is quite natural from the point of view of optics. We reflect the point A and the lines AC and AD in the mirror g (Fig. 15). If A' is the image of A, then $A'C$ is the image of AC and $A'D$ is the image of AD, so that $A'C = AC$, $A'D = AD$, and $A'E = AE$. Therefore the triangles EDA and EDA' are congruent and the angles EDA and EDA' are equal. According to our hypothesis we have angle EDA = angle CDB.

27

Therefore angles *CDB* and *EDA'* play the roles of vertical angles; that is, *A'DB* is a straight line.

Now the lengths of the paths *ADB* and *A'DB*, as well as *ACB* and *A'CB*, are equal. Since *A'DB* is a straight line connecting *A'* and *B*, it is shorter than the path *A'CB*, and consequently *ADB* is shorter than *ACB*. Here we have used the fact that a straight line is the shortest distance between two points.

2. We now turn to our main problem, *to inscribe in a given acute-angle triangle ABC a triangle UVW whose perimeter is as small as possible* (Fig. 16). *The assertion is that the "pedal triangle" EFG* (Fig. 17), *whose vertices are the three feet of the altitudes of the triangle ABC, has a smaller perimeter than any other inscribed triangle UVW.*

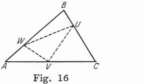
Fig. 16

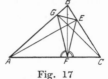
Fig. 17

We must first prove a lemma concerning the pedal triangle. We assert the angles *AFG* and *CFE* are equal (as in the law of reflection), and consequently that the analogous angles at *E* are equal and also those at *G*. In order to prove this lemma we must recollect some theorems of plane geometry: the theorem of Thales, that an angle inscribed in a semicircle is a right angle (Fig. 18); that inscribed angles which intercept the same arc are equal (Fig. 19); that the altitudes of a triangle meet at a point. Using these, we see that the circle with diameter *AH* passes through *G* and *F* and that the circle with

Fig. 18

Fig. 19

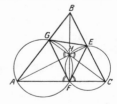
Fig. 20

diameter *CH* passes through *E* and F (Fig. 20). Furthermore, angle *AFG* intercepts the arc *AG*, as does angle *AHG*, so these two angles are equal. In the same way we see that the angles *CFE* and *CHE* are also equal. But angles *AHG* and *CHE* are vertical angles and hence are equal. Therefore we have angle *AFG* = angle *CFE*.

3. We can now commence Schwarz's proof. We reflect the triangle *ABC* in the side *BC* (Fig. 21), reflect the reflected triangle in its side *CA'*, reflect the resulting triangle in *A'B'*, reflect in *B'C'*, then *C'A''*, then *A''B''*, a total of six reflections. We first prove the fairly obvious fact that the final position *A''B''C''* is the original position *ABC* moved parallel to itself without turning. The first two reflections shift *ABC* to the third position *A'B'C*. This shift could have been made without reflections and without lifting the triangle out of its plane merely by rotating it about *C* through the angle 2*C* in a clockwise direction. Similarly, the shift from the third to the fifth position could be accomplished by a clockwise rotation through the angle 2*B* about the point *B'*. Finally, a clockwise rotation through the angle 2*A* about *A''* would yield the seventh or final position. In all, the triangle has been rotated one complete revolution, through the angle 2*C* + 2*B* + 2*A*, since the sum *A* + *B* + *C* of the angles of a triangle is a straight angle. The final position of the triangle therefore has the same orientation as the original; it has merely been moved but remains parallel to itself. *Therefore BC is parallel to B''C''.*

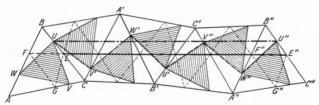

Fig. 21

Now we want to trace the various positions assumed by the pedal triangle and the triangle *UVW* under the successive reflections. These are shown in Fig. 21 by dotted lines and shading. From our lemma concerning the pedal triangle, we see at once that the second position of *EG* forms a straight line with the first position of *FE*. In the same way, one side of the pedal triangle will always lie on the continuation of this line in successive positions. Therefore *the straight line EE''* is made up of 6 segments, 2 equal to *FG*, 2 to *GE*, and 2 to *EF*: hence it *is equal to twice the perimeter of the pedal triangle.*

Tracing out the positions assumed by the arbitrary triangle *UVW* in the same way, we find that *the zig-zag line UV'W'U'V''W''U''*, *connecting U and U'', is equal to twice the perimeter of the triangle UVW.*

We have already seen that the segments *UE* and *U''E''* are parallel since they lie on *BC* and *B''C''*. They are also equal, since

29

they are corresponding segments in two positions of the triangle
ABC. Then by a theorem of plane geometry $EE''UU''$ is a parallelogram, and consequently its other two sides are equal, $UU''=EE''$.
Therefore UU'' is also equal to twice the perimeter of the pedal
triangle. The straight line UU'' connecting U and U'' is shorter
than the zig-zag line connecting the same two points, and the zig-zag
line is twice the perimeter of UVW. Therefore the perimeter of
the pedal triangle is less than the perimeter of UVW, as was to be
proved.

This proof is typical of many truly mathematical proofs. It
consists essentially in transforming the hypotheses and conclusion
until the true kernel of the theorem can be recognized at a glance.

6. A Second Proof of the Same Minimum Property

1. In the last chapter we proved that of all the triangles inscribed
in a given acute-angle triangle, the pedal triangle has the smallest
perimeter. It is worthwhile to consider another proof of the same
theorem, because this second proof will illustrate some new ideas
and, for our purposes, the methods used are of more importance and
interest than the mere mathematical content of new theorems.
The previous proof, originally given by H. A. Schwarz, depended
essentially on the fact that a straight line is the shortest distance
between two points, and it made use of the idea of reflection of a
figure in a line. These two principles also form the basis of the
second proof, and it is of interest to contrast the manner in which
they are used in the two proofs. The following proof was given by
L. Fejér, who discovered it as a student and thereby won especial
recognition from H. A. Schwarz.

2. In the given acute-angle triangle ABC (Fig. 22), let UVW be
an arbitrary inscribed triangle with U on BC, V on CA, W on AB.

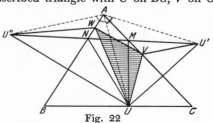

Fig. 22

Let U be reflected in the two lines AC and AB and call U' and U''
the two images. Now UV and $U'V$ are mirror images of each other
and hence are equal. For the same reason UW and $U''W$ are also

equal. The perimeter of the triangle UVW is $UV + VW + WU$ and therefore it is equal to the length of the path $U'VWU''$.

If we hold U fixed but move V and W to new positions, then the points U' and U'' remain fixed because they depend only on U and the triangle ABC. The path $U'VWU''$ then connects the two fixed points U' and U'', and its length is always equal to the perimeter of UVW. The shortest path from U' to U'' is a straight line. Therefore the straight line segment $U'U''$ is the smallest possible perimeter for an inscribed triangle with one vertex held at U. This minimal triangle with vertex at U, which we shall call UMN, is shown in Fig. 22.

3. Having found the triangle of smallest perimeter with vertex at U, we need only compare the minimal triangles for various positions of U and pick out the one with smallest perimeter. That triangle will certainly have the smallest perimeter of all inscribed triangles.

We must determine the position of U so that the segment $U'U''$ is as small as possible. For this purpose we first notice that the triangle $AU'U''$ is isosceles, with AU' and AU'' as equal sides. Indeed these two segments are each mirror images of the same segment AU and hence are equal, $AU = AU' = AU''$.

Although the legs of the triangle $AU'U''$ are equal in length to AU and hence depend on the position of U on BC, *the size of the angle $U''AU'$ does not depend on the position of U.* This angle is completely determined by the original triangle ABC and nothing else, since (because of the reflections) we have the following equations between the angles of the figure:

$$UAB = U''AB, \quad UAC = U'AC.$$

From the first we have

$$U''AU = 2UAB,$$

and from the second,

$$U'AU = 2UAC;$$

and hence

$$U'AU + U''AU = 2UAB + 2UAC$$

or

$$U'AU'' = 2BAC,$$

which proves our assertion concerning the angle $U'AU''$.

4. In the isosceles triangle $AU'U''$ we want to make the base $U'U''$ as small as possible. Since the angle at A does not depend on U, all these triangles, for different positions of U, have the same

vertex angle. Of all these the one with the shortest base will also have the shortest legs. The legs AU' and AU'' have the length AU. Therefore we will obtain the shortest segment $U'U''$ if we choose U so that AU is as short as possible.

Now the segment AU connects the point A with the line BC, and it is well known that the shortest distance from a point to a line is the perpendicular distance. Therefore we must choose U so that AU is perpendicular to BC, that is, so that AU is the altitude of the triangle ABC through A.

5. Let us now construct this triangle EFG of smallest perimeter (Fig. 23). Let E be the foot of the perpendicular to BC through A. If E' and E'' are the images of E under reflection in AC and AB, then $E'E''$ is the length of the smallest perimeter of an inscribed triangle. The two points F and G where the line $E'E''$ cuts AC and AB are the other two vertices of the minimal triangle.

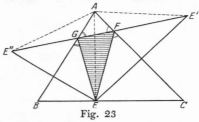

Fig. 23

If we think back over what we have done, we see that every inscribed triangle UVW, different from EFG, must have a larger perimeter. For if U is different from E, then the segment $U'U''$ is larger than $E'E''$ and the perimeter of UVW is at least as large as $U'U''$. If U and E coincide, then one or both of the points V and W will differ from the points F and G, and the path $E'VWE''$ will differ from the straight line $E'FGE''$. In both cases, then, the perimeter of UVW will actually be greater than the perimeter of EFG.

6. These considerations have shown that the problem of finding an inscribed triangle of least possible perimeter has only one solution. We shall make use of this uniqueness of the solution. Our construction of the minimal triangle did not treat all three vertices in the same manner. One vertex E is the foot of the altitude through A, but the other two were found by a construction having nothing to do with the altitudes through B and C.

We could have carried out our argument starting with the vertex B in place of A. That is, in § 2, instead of reflecting the point U in the sides AB and AC, we could have reflected the point V in BA

and BC and continued accordingly. We would then have ended with a minimal triangle whose vertex F was the foot of the altitude through B. Since there is only *one* minimal triangle, this construction starting with B must have led to exactly the same triangle EFG as the original construction starting with A. Since we could also start with the vertex C, we may conclude that in the minimal triangle EFG not only E but F and G as well are feet of altitudes. Thus we have proved the theorem.

7. We can still get a little more from the proof. Making use of the uniqueness of the solution, we saw that if E had a certain property (being a foot of an altitude), then F and G also had the analogous property. Similarly, any property that F and G have by our construction will also hold for E. Because E' is the mirror image of E, the angles EFC and $E'FC$ are equal. Since $E'FC$ and GFA are vertical angles, they are also equal, and angle $EFC =$ angle GFA. That is, the two sides of the minimal triangle that go through F form equal angles with the side AC of the original triangle. The corresponding statement holds for the point G. If we had started our construction with F as the foot of the altitude through B, then this same proof would show that the angles GEB and FEC are also equal.

Disregarding the minimal property of the triangle EFG, we know from § 6 that EFG can be characterized as the pedal triangle. Combining these two results, we have the theorem: if a pedal triangle is inscribed in an acute-angle traingle, then the two angles formed at each vertex of the pedal triangle by the two sides of the pedal triangle and the side of the original triangle are equal.

This theorem now contains nothing concerning a minimum. It is the type of theorem customarily found in elementary geometry, and it could be proved by the methods of elementary geometry. In fact, we have actually done just that in Chapter 5. Schwarz's proof needed this result as a lemma, and we supplied it by making use of a number of theorems concerning circles. An advantage of Fejér's proof is that it makes use of nothing other than the principle of the shortest distance and reflections. Furthermore, Fejér's proof is distinguished by the fact that it uses only two reflections, while Schwarz's employs six.

8. There is a counterpart to the theorem on the pedal triangle: In every acute-angle triangle there is one and only one point the sum of whose distances from the three vertices is a minimum. This point is so situated that the lines that join it to the three vertices form angles of $120°$ with each other.

This theorem was proved by L. Schruttka by a method suggested by analogy with Schwarz's proof of the pedal triangle theorem. A much shorter proof has been given by Bückner, and this is the one we shall use.

Let P (Fig. 24a) be an arbitrary point in the acute-angle triangle ABC. Let the triangle ACP be turned about the point A through $60°$ to the position $AC'P'$. This rotation is to be made in such a direction that AC turns out of the triangle, so that finally the line

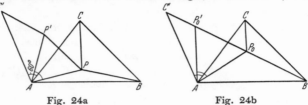

Fig. 24a   Fig. 24b

AC lies between AB and AC'. Then we have $C'P' = CP$ and $PP' = AP$, for the triangle APP' is not only isosceles but equilateral, its angle at A being $60°$. *Then the path $BPP'C'$ represents the sum of the distances of P from the three vertices A, B, C.* The point C' is independent of the position of P. All the paths corresponding to various positions of P join B and C'. The shortest of these paths is the straight line BC' (Fig. 24b). Therefore the minimal point P_0 must lie on BC', and its position is completely determined by the fact that angle $AP_0C' = 60°$. The supplementary angle AP_0B is then $120°$. The construction shows that there can be only *one* minimal point P_0. Consequently the same construction with A replaced by another vertex will lead to the same point P_0. Therefore the angles BP_0C and CP_0A are also $120°$.

7. The Theory of Sets

The subject of this chapter lies at the very foundations of mathematics. However, our interest in it will depend more on the beauty and simplicity of the manner in which it is built up than on its significance for mathematics in general. The theory of sets, originated by Georg Cantor, is a truly mathematical theory which starts with only the very simplest concepts and builds up to a ramified and important subject through the use of pure reasoning.

Are there more whole numbers than even numbers? Which are more numerous, the points of a line segment or the points of the surface of a square? It is just such questions as these that started

Cantor on his theory. It is important to avoid jumping to conclusions when trying to answer these questions. The questions as they are put are not precise because we don't know exactly what we mean by one being more than another. Cantor's first important step was to give them a precise meaning by using simple methods of counting off, as is done for finite numbers, and by carefully differentiating between cardinal and ordinal numbers, a distinction that is merely grammatical for finite numbers.

A simple example will point out the direction in which we are to go. Suppose we are in a dance hall and are asked whether there are more men or women present. What would be the easiest way to decide? One method would be to separate the men and women into two groups, to count each group, and to compare the numbers. A simpler method, however, would be to start the dance. The men and women would pair off, and it would only be necessary to observe whether those left over were men or women. We suppose that everyone dances if he can find a partner.

This principle of pairing off was adopted by Cantor as a starting point. If we wish to decide whether there are more whole numbers than even numbers we are to try to pair them off. In fact, we can find a pairing in which each whole number is paired with an even number and none is left out. We do it as follows:

$$1, \ 2, \ 3, \ 4, \ 5, \ 6, \ \cdots$$
$$2, \ 4, \ 6, \ 8, \ 10, \ 12, \cdots$$

where each whole number in the upper row is to be paired off with the even number directly below it. By this method every number gets paired off and none is left out. This simple fact is quite remarkable. The two rows have been paired off exactly and yet the lower row consists of part and only part of the upper row.

This brings up an essential difference between this case and the case of finite numbers. In the case of the dance hall (where there is a finite number of dancers) it is quite immaterial which man dances with which woman. The number of those sitting out is always the same and it does not change as long as no one enters or leaves the hall. It is quite different in the case of the whole numbers and even numbers. We have seen a pairing off that comes out exactly, including every element, but it is easy to give another which does not do so. We can use the most natural pairing: 2 with 2, 4 with 4, 6 with 6, etc. We have paired off the even whole numbers, but all of the odd ones are left over. The essence of Cantor's theory is that it abandons the idea of arbitrary pairing and only demands that we find some *one* pairing off that comes out exactly. If such a pairing can be found, the two sets are said to be *of the same power*.

Cantor's next step ˙was to prove that *the set of all rational numbers, i.e. all whole numbers and fractions, does not have higher power than the set of whole numbers.* To construct the appropriate pairing we arrange the fractions not in order of size but according to the size of the sum of their numerator and denominator. We consider only reduced fractions and begin with those the sum of whose numerator and denominator is 2. There is only one such fraction: $1/1 = 1$. Those with sum 3 are $1/2$ and $2/1 = 2$. Next come $1/3$, $2/2$, $3/1$, with sum 4, but $2/2$ is omitted because it is not reduced. These are then followed by $1/4$, $2/3$, $3/2$, $4/1$, etc. We now write the fractions in this order under the series of natural numbers,

$$1, \quad 2, \quad 3, \quad 4, \quad 5, \quad 6, \quad 7, \quad 8, \quad 9, \quad 10, \quad 11, \quad 12, \quad \cdots$$
$$1, \quad 1/2, \quad 2, \quad 1/3, \quad 3, \quad 1/4, \quad 2/3, \quad 3/2, \quad 4, \quad 1/5, \quad 5, \quad 1/6, \cdots,$$

and can pair off each whole number with the fraction directly under it. The lower row does not leave out any rational number, because each one has a definite sum of numerator and denominator and must therefore have some place in the row. This, then, is the pairing off we desired.

This surprising result is expressed by saying that the set of rational numbers is "countable" or "denumerable" because the pairing off with the natural numbers is, in effect, a counting off of the rationals. In general, a set is denumerable when it can be paired off with the natural numbers, that is, when it has the same power as the set of the natural numbers. Cantor goes on to show that several other sets which are much more extensive than the natural numbers do not actually have higher power. We shall pass by these examples since they involve further mathematical concepts and our one example is adequate.

Thus far all the infinite sets considered have had the same power. Cantor's theory would be trivial if there were no sets of higher power. Cantor showed that sets of higher power do exist by proving that *the set of points of a line segment is of higher power than the set of natural numbers.* The proof is an indirect one. We suppose that there is a pairing between the points of, say, a 1 inch segment and the natural numbers, and show that this leads to a contradiction. This pairing off puts the points in an order, the first paired with 1, the second with 2, etc., but this will obviously not be the natural order of the points on the line. It is convenient to identify each point by its distance from one end of the line segment. For example, the middle of the segment corresponds to the number 0.5 and each point corresponds to some decimal fraction. In order to measure the points exactly we must use infinite decimals; for example the end of the first one-third of the segment corresponds to $0.33333 \cdots$. In the pairing off of the points of the segment with the natural numbers we can use the corresponding infinite decimal fractions instead of the points themselves. In the first place there will be some decimal $0. \cdots$ corresponding to 1, in the second place a similar decimal, etc. It is a little easier to list them in a vertical column, giving us a list of the type shown below where we have inserted particular numbers merely for purposes of illustration.

1. $0.35420 \cdots$
2. $0.61773 \cdots$
3. $0.55549 \cdots$
4. $0.01007 \cdots$
5. $0.20206 \cdots$

  .   . . .
  .   . . .

Now we shall show that there is a decimal fraction $0. \cdots$ which, contrary to our assumption, is not included in the list. This decimal can be found as follows: we choose for the first digit after the decimal point a digit that is different from the first digit of the first decimal in the list. This gives us a choice of 9 digits. In order to make it definite, let us choose the digit 1 unless the first digit of the first decimal in the list is 1, in which case we shall choose 2. Now it is clear that our new decimal will differ from the first decimal of the list no matter what choice is made for the remaining digits, for the two certainly differ in their first digits and will represent different points even if they agree in all the other digits. We now choose 1 to be the second digit unless (as in the above example) the second digit of the second decimal of the list is 1, in which case we choose 2. Therefore in either case the second digit of our decimal is different from the second digit of the second decimal of the list. Then our decimal will differ from the second decimal of the list as well as from the first. We continue choosing digits in this way. In the illustrative example our decimal would begin $0.12111 \cdots$. Since the process can be continued indefinitely, we have defined an infinite decimal fraction which is different from all the decimals in the list. The fact that the list does not contain this decimal contradicts the assumption that they could be paired off. Therefore the points of a line segment are not denumerable.

There is a certain objection to the above proof which must be mentioned. It is caused by the decimals that end in an infinite row of 9's, like $0.269999\cdots$, which is no different from $0.270000\cdots = 0.27$. The difficulty is that two different decimals can represent the same point if one of them ends in 9's, and this means that we were not justified in replacing the points of the segment by the set of all decimal fractions.[3] The two sets do not correspond exactly. Cantor avoided this difficulty by putting the proof in quite a different form, but there is a very simple way to dispose of the objection. We merely rule out all decimals ending in 9's. That is, we would not identify a point by $0.269999\cdots$ but would use $0.270000\cdots$. All that we have to do is to make sure that the decimal we construct does not end in 9's. But this does not occur, since it is made up entirely of the digits 1 and 2 and does not contain any 9's.

This result has an interesting consequence. Since the set of rational numbers is denumerable, we see that it is of lower power than the set of all numbers between 0 and 1. Therefore there must certainly be numbers between 0 and 1 which are not rational. Thus we have proved the existence of irrational numbers by means of quite general considerations. The same thing was proved in an entirely different way in chapter 4.

Cantor's next result is also rather surprising. It states that *the set of points of the surface of a square does not have higher power than the set of points of a side of the square.* The reason this is so surprising is that it contradicts the intuitive idea of dimension. A one-dimensional segment has the same power as a two-dimensional square, and it can also be shown that a three-dimensional cube also has the same power.

As in the previous proof we shall use infinite decimal fractions and again we shall exclude decimals ending in 9's. The points of the segment will again be represented by decimals $0.\cdots$ just as before. The points of the square will be characterized by a pair of decimals, one decimal, x, representing the horizontal distance from the left side of the square and another, y, representing the vertical distance from the lower side of the square (Fig. 25). We shall

[3] For a further discussion, see R. Courant and H. Robbins, *What is Mathematics?*, 1941, Oxford University Press, New York, pp. 64–66, 80–82.

set up a pairing in the following way. Each point P of the square is characterized by two decimals

Fig. 25

$$x = 0.a_1a_2a_3 \cdots, \quad y = 0.b_1b_2b_3 \cdots.$$

From these two we form the single decimal

$$z = 0.a_1b_1a_2b_2a_3b_3 \cdots$$

by alternately taking digits from x and y. Then we pair off the point P of the square with the point Q of the segment characterized by the decimal z. For example, the center of the square has $x = 0.500 \cdots$, $y = 0.500 \cdots$, and it is paired off with the point of the segment corresponding to $z = 0.550000 \cdots$. In this way each point of the square is paired off with some point of the segment. But this is not quite enough. If we had only to pair each point of the square with some point of the segment we could merely pair each point P with the point directly above P on the upper side of the square. But then each point of the side would be paired, not with just *one* point of the square, but with an infinite number (all the points on a vertical line). But this is ruled out in Cantor's method of comparing powers of sets, just as in the case of the dancers we did not suppose that one person would dance with several partners. Our original, more refined, method of pairing off avoids this difficulty. For if a second point P' of the square with, say,

$$x' = 0.a_1'a_2'a_3' \cdots, \quad y' = 0.b_1'b_2'b_3' \cdots$$

was paired with the same point Q of the segment, then we would have

$$z = 0.a_1'b_1'a_2'b_2'a_3'b_3' \cdots.$$

This decimal expression for z and the original one can be equal only if the two correspond exactly in all digits,

$$a_1' = a_1, \quad b_1' = b_1, \quad a_2' = a_2, \quad b_2' = b_2, \cdots.$$

But this shows that all the digits of x' are equal to the corresponding digits of x, and the same is true for y' and y. Therefore we have

39

$x' = x$, $y' = y$, and P' is the same point as P, contrary to the assumption that it was a different point. Consequently two different points of the square are never paired off with the same point of the segment.

The pairing will be complete when we show that no point of the segment is left unpaired. This is easily seen, for if

$$z = 0.c_1c_2c_3c_4c_5c_6 \cdots$$

corresponds to any point of the segment, then the point of the square with

$$x = 0.c_1c_3c_5 \cdots, \quad y = 0.c_2c_4c_6 \cdots$$

is paired off with the point corresponding to the decimal obtained by taking digits alternately from x and y, and this is exactly the decimal z with which we started.

There is still an objection to this proof which arises from decimals ending in 9's. The point on the segment with $z = 0.2202020 \cdots$ is paired with the point of the square having

$$x = 0.2000 \cdots, \quad y = 0.2222 \cdots.$$

The point on the segment with $z' = 0.12929292 \cdots$ is paired with the point of the square having

$$x' = 0.1999 \cdots, \quad y' = 0.2222 \cdots,$$

but here we have a decimal ending in 9's and we should write $x' = 0.2000 \cdots$. Therefore the two different points on the segment, corresponding to z and z', are paired with the same point of the square.

A very simple, although not obvious, trick will eliminate this difficulty. In the exact form in which it is given above, the proof is not correct, but if we make the one following modification it becomes valid. Whenever a digit 9 occurs in a decimal, we combine it with the digit to its right to form an inseparable "molecule". If there are several consecutive 9's we include all of them and the next digit in the molecule. Thus in the above example we would have $0.(1)(2)(92)(92)(92) \cdots$ and, as a new example, we would write $z'' = 0.(7)(3)(94)(990)(9997) \cdots$. These molecules are to take the place of the digits in the proof, so that from z'', for example, we would have

$$x'' = 0.(7)(94)(9997) \cdots, \quad y'' = 0.(3)(990) \cdots.$$

The pairing is now quite different from what it was before but, after a moment's reflection, we see that the proof now goes through without difficult so long as we remember to rule out decimals ending in 9's.

At this point Cantor encountered a problem which has since become famous. This was the question whether there exists a set having a higher power than the set of natural numbers but a lower power than the set of points of a segment.

This problem has been named "the problem of the continuum". It defied all of Cantor's attempts to find an answer, as well as those of his successors. Probably no other mathematical problem formulated with so little mathematical preparation has ever defied solution so stubbornly as this one. The only concepts used in the formulation of the problem are those of whole numbers and line segments. There is nothing remarkable in proposing one or many very difficult problems if one uses complicated mathematical ideas. Using only simple ideas to formulate a problem that is neither easily solved nor trivial reveals the true art of mathematics. Considered from this point of view, the problem of the continuum stands out as a shining example.

It soon became evident that the fundamental theory of sets, on which all of mathematics depends, could not be properly developed until the concept of what is meant by a set had been carefully analyzed. One is forced to contemplate such an analysis because of the following famous *paradox* in the theory of sets.

We have been talking of sets. These sets consist of "elements" of one sort or another. The set of points of a segment contains as elements the individual points of the segment. The whole numbers themselves are the elements of the set of whole numbers. The relation between the sets and elements is the same as that between an association and its members. Sometimes an association has as members not individuals but associations. For example, the United Nations is an association of nations, each of which is an association of members or citizens. The actual membership of the United Nations is made up of individual countries, not of the citizens of those countries. In the same way a set can contain sets as elements. An example is the set of all denumerable sets, all of whose elements are sets in themselves. Being a citizen of a participating nation does not make a person a member of the United Nations. Similarly the number 1/5 is an element of the set of rational numbers, which we have proved denumerable, but that does not make it an element of the set of all denumerable sets.

41

Can a set contain itself as an element? The ordinary sets which we have been considering do not have this property. However, it is easy to give examples to show that such extraordinary sets do exist. The set of all conceivable sets is of this type, since it is a particular set in itself. For the moment we shall call a set that contains itself as an element an extraordinary set, while we shall call the others ordinary.

Now let us consider the set of all ordinary sets, and let us call it the set *s*. Is *s* itself an ordinary or an extraordinary set? It must be one or the other. If *s* is extraordinary then it must contain itself as an element. But then *s* is a member of *s* and hence is an ordinary set, as are all elements of *s*. This is a contradiction, so *s* is not extraordinary. However, if *s* is ordinary, then it does not contain itself as an element. Therefore *s* is not an element of *s*, which contains all ordinary sets as elements, and so *s* is not ordinary. This is again a contradiction, so *s* is not ordinary. This is the paradox: *s* must be either ordinary or extraordinary, but each possibility leads to a contradiction.

This paradox is not specifically restricted to the theory of sets. In order to make this clear we shall reformulate the same paradox in a rather frivolous way. This formulation is completely free of the idea of sets. In a certain regiment a soldier is detailed to take over the duties of barber. His exact orders direct him to shave everyone in the regiment who does not shave himself. Should the soldier shave himself? If he shaves himself, then he is one who shaves himself, and his orders direct him not to. If he doesn't shave himself then he is one who does not shave himself and again he violates his orders. What can the man do to carry out his orders strictly?

The paradox is a purely logical one. We are inescapably led to it in the theory of sets, but it is more general than that and it does not need that theory for its formulation. The old, rather dull, subject of logic has developed into something quite interesting. As a matter of fact, both mathematicians and logicians have been working for some time to free logic from its old Aristotelian form, but at present it is difficult to see just what new form it will take.

8. Some Combinatorial Problems

1. A simple example will serve to show what type of problem we shall discuss. Suppose we have 4 red balls (R), 1 yellow ball (Y), and 2 white balls (W). These balls are supposed to be of the same size and weight and completely indistinguishable except by their colors. We also suppose that we have two urns, A and B. Urn A will hold exactly 3 balls, B will hold 4. *In how many different ways can the 7 colored balls be distributed between the urns A and B?*

Since we have a very simple case with only two urns, we need only consider the balls that are put into A. The remaining 4 balls will then have to be put into B. To answer our question we shall systematically list all the possibilities. First of all, A might contain only red balls and B whatever is left over:

1.   in $A : RRR$, in $B : RYWW$.

It does not matter which particular three red balls are put into A, since we supposed that the balls could not be distinguished from each other except by their colors.

Next A might contain just 2 red balls, in which case the third ball must be either yellow or white. This gives the two distributions

2.   in $A : RRY$, in $B : RRWW$,
3.   in $A : RRW$, in $B : RRYW$.

If A contains just 1 red ball, then the other two will clearly have to be either YW or WW, giving

4.   in $A : RYW$, in $B : RRRW$,
5.   in $A : RWW$, in $B : RRRY$.

Finally, if A contains no red balls, it must contain the other three:

6.   in $A : YWW$, in $B : RRRR$.

Therefore we see that there are 6 possible ways to distribute 4 red, 1 yellow, and 2 white balls between two urns which hold 3 and 4 balls respectively. Clearly it is quite immaterial of what colors the balls are; we would get the same result with 4 black, 1 green, and 2 blue balls, so it is necessary only to tell the number of balls of each color, not their colors. In order to state our result a little more briefly, we will then say: the number of distributions of 4, 1, 2, balls between urns of contents 3 and 4 is 6. We will also write the same fact in symbols in the form:

$$\{4,\ 1,\ 2\ |\ 3,\ 4\}_7 = 6.$$

Here the numbers in front of the vertical bar are the number of balls of the various colors, and the numbers after the bar are the number of balls that each urn can hold. *The sum of the numbers in front of the bar must equal the sum of the numbers after the bar*, since all the balls together just fill all the urns. This total number of balls, which is 7 in this case, has been written as a subscript to the right of the bracket.

2. It is not necessary to consider just three colors and two urns. As a general problem we might consider n balls of c different colors and u urns of total content n. We would then use the symbol

$$(1) \qquad Z = \{r, s, \cdots \mid a, b, \cdots\}_n$$

for the number of distributions of n balls, of which r are one color, s another, etc., among urns which will hold a, b, $\cdots$ balls respectively. The problem would be to compute the value of Z from the numbers n, r, s, $\cdots$, a, b, $\cdots$. We will not solve this problem in such a general form, but rather will restrict ourselves to a number of examples and important special cases.

The problem does not really require the objects distributed to be colored balls. For example, we have already used the letters *RRRRYWW* to designate the 4 red, 1 yellow, and 2 white balls, and have distributed these 7 *letters* between 2 *sets A* and *B* instead of distributing 7 *balls* between 2 *urns A* and *B*.

3. We shall take up a number of examples having nothing to do with colored balls, but we will try to discover a way of interpreting each case as a distribution of colored balls. These examples are of considerable interest and importance and they will serve to show the significance of the symbol (1).

Example I. In how many ways can n persons be seated at n places? In order to interpret this in the form of our previous problem, we note that each person can be distinguished from every other one. Therefore we can designate each person by a different color, a name, so to speak. The problem then reduces to: in how many ways can n balls of n *different* colors be put in n different places? Each of these n places corresponds in our old problem to an urn that will hold just one ball. Using our symbol, the problem reduces to finding the value of

$$(2) \qquad P_n = \{1, 1, \cdots \mid 1, 1, \cdots\}_n$$

where there are n 1's in front of the vertical bar, corresponding to n different colored balls, and n 1's after the bar corresponding to the n urns, each holding just 1 ball.

If we think of the n urns or places as being put in a row, then we are asking in how many ways n balls or persons (or any distinguishable objects) can be arranged in a row? Such an arrangement is called a *permutation*, so we can say that P_n of (2) represents the number of permutations of n different things.

There still remains the problem of finding the numerical value of P_n if we are told how big n is. We shall find a formula for P_n in terms of n, but shall postpone it until we have taken up a number of further examples.

4. *Example II.* In the game of skat 32 different cards are used and 3 players participate. Each player is dealt 10 cards and the remaining 2 cards go into the "skat". In how many different ways can the hands be dealt out? Clearly the number of ways is the same as the number of distributions of 32 differently colored balls among 4 urns that will hold 10, 10, 10, 2 balls respectively. Therefore the number of ways the cards can be distributed in a game of skat is given by

$$(3) \qquad S = \{1, 1, \cdots \mid 10, 10, 10, 2\}_{32}.$$

Again we will postpone the numerical computation of this symbol.

5. *Example III.* The so-called *polynomial theorem* is another interesting example. We shall consider only a special case involving three variables x, y, z. The expression

$$(x + y + z)^n$$

is to be multiplied out. This n-th power represents a product of n equal factors, each of which is $(x + y + z)$. If a sum of terms enclosed in parentheses is to be multiplied by a factor, then each term inside the parentheses must be multiplied by that factor. If this rule is applied to all the n parentheses in the n-th power, then it is seen that the power consists of a sum of products. Each of these products has n factors, which can be x or y or z. Since the factors of a product can be taken in any order we can bring all the x's together, then the y's, and then the z's. Then each product will be of the form $x^a y^b z^c$ where a, b, c, are whole numbers subject to the restrictions

$$(4) \qquad a + b + c = n,\ a \geqq 0,\ b \geqq 0,\ c \geqq 0.$$

A factor x can originate in any of the n parentheses and so can factors y and z. Therefore there can be several products $x^a y^b z^c$ with the same values of a, b, c. How often will $x^a y^b z^c$ arise when we multiply out the n-th power? One of the factors x, y, or z must originate from each of the n parentheses. We can imagine a row of n urns, one corresponding to each parenthesis. Then we can put into each urn the letter that originates in the corresponding parenthesis. Now we need only count how many ways a elements "x" (red balls), b elements "y" (yellow balls), and c elements "z" (white balls) can be distributed among n urns each holding just one element. We shall designate this number by $P_{a,b,c}^{(n)}$ and we now have

(5) $$P_{a,b,c}^{(n)} = \{a, b, c \mid 1, 1, \cdots, 1\}_n.$$

When the power of $(x + y + z)^n$ is multiplied out, the term $x^a y^b z^c$ will occur $P_{a,b,c}^{(n)}$ times. These like terms can be collected, and the combined term will then have the coefficient $P_{a,b,c}^{(n)}$.

A particular case will help to make this result clearer. Let us take $n = 4$ and think of multiplying out the power $(x + y + z)^4$. Terms of the type $x^a y^b z^c$ will occur with all values of a, b, c that are consistent (4) with $n = 4$. Listing them systematically, we find the following 15 possibilities:

$$
\begin{array}{llllll}
x^4, & y^4, & z^4, \\
x^3 y, & x^3 z, & xy^3, & y^3 z, & xz^3, & yz^3. \\
x^2 y^2, & x^2 z^2, & y^2 z^2, \\
x^2 yz, & xy^2 z, & xyz^2,
\end{array}
$$

Supplying these terms with the coefficients we have found, we get

(6)
$$
\begin{aligned}
(x + y + z)^4 = & \; P_{4,0,0}^{(4)} x^4 + P_{0,4,0}^{(4)} y^4 + P_{0,0,4}^{(4)} z^4 \\
& + P_{3,1,0}^{(4)} x^3 y + P_{3,0,1}^{(4)} x^3 z + P_{1,3,0}^{(4)} xy^3 + P_{0,3,1}^{(4)} y^3 z \\
& \qquad\qquad + P_{1,0,3}^{(4)} xz^3 + P_{0,1,3}^{(4)} yz^3 \\
& + P_{2,2,0}^{(4)} x^2 y^2 + P_{2,0,2}^{(4)} x^2 z^2 + P_{0,2,2}^{(4)} y^2 z^2 \\
& + P_{2,1,1}^{(4)} x^2 yz + P_{1,2,1}^{(4)} xy^2 z + P_{1,1,2}^{(4)} xyz^2.
\end{aligned}
$$

This gives only the *form* of the power. We must still find the values of $P_{a,b,c}^{(4)}$ and, more generally, of $P_{a,b,c}^{(n)}$.

6. *Example IV.* In how many different ways can k things be chosen from among n different things? This number is usually called the number of combinations of n things taken k at a time and is designated by the symbol $C_k^{(n)}$. The n things are all different, so we can consider them as differently colored balls, $r = 1$, $s = 1$, $\cdots$. The k balls that are chosen can be put in one urn, $a = k$, and the remaining ones can be put in another urn, $b = n - k$. The number we are trying to determine is therefore

(7) $$C_k^{(n)} = \{1, 1, \cdots \mid k, n - k\}_n.$$

7. *The duality of the distribution symbol.* The symbol that we have used to represent the number of distributions has a very important property that will help us to compute its value. The numbers in front of the vertical bar and those following it can be interchanged,

$$(8) \qquad \{r, s, \cdots \mid a, b, \cdots\}_n = \{a, b, \cdots \mid r, s, \cdots\}_n.$$

Stated in terms of distributions, this asserts that there are *exactly the same number* of ways to distribute r red balls, s white, $\cdots$ among urns that will hold a, b, $\cdots$ balls respectively, as there are ways to distribute a red balls, b white, $\cdots$ among urns that will hold r, s, $\cdots$ balls respectively. Here, as always, we are supposing that $r + s + \cdots = a + b + \cdots = n$.

Expressed in this way, in terms of distributions, the equality (8) is very easily proved. A simple numerical example will completely demonstrate the proof. Let us prove the equality

$$(9) \qquad \{3, 4 \mid 1, 1, 5\}_7 = \{1, 1, 5 \mid 3, 4\}_7.$$

The left side represents the number of distributions of 3 red and 4 white balls

$$R, R, R, W, W, W, W,$$

among the three urns, A and B each holding 1 ball and C holding the other 5. Let us look at one of the possible distributions, say

$$\underline{|R|} \qquad \underline{|W|} \qquad \underline{|RRWWW|}$$
$$A \qquad B \qquad C$$

Now we can just as well write this distribution by listing the balls in a row and putting under each one the urn into which it goes,

$$(10a) \qquad \begin{matrix} R, & W, & R, & R, & W, & W, & W \\ A, & B, & C, & C, & C, & C, & C. \end{matrix}$$

This is just a list of 7 pairs of letters. The upper letters R and W are the names of the colors, the lower letters A, B, C are the names of the urns. We can interchange the roles of the letters and let A, B, C be the names of colors, R and W the names of urns. If we now write the colors in the upper row, the urns in the lower, and rearrange the pairs according to urns, we have the scheme

$$(10b) \qquad \begin{matrix} A, & C, & C, & B, & C, & C, & C \\ R, & R, & R, & W, & W, & W, & W. \end{matrix}$$

Exactly the same pairs appear in (10b) as in (10a), only each has been inverted. But (10b) can be interpreted as representing a distribution of 1 ball of color A, 1 of B, and 5 of C among the two urns R and W, of content 3 and 4 respectively. But this is one of the distributions which are counted by the right side of (9). Every other distribution counted by the left side of (9) will correspond to one counted by the right side, in the same way as (10a) corresponds to (10b). This correspondence clearly works in the other direction, from (10b) to (10a), in just the same way. Therefore there is a complete correspondence, a "duality," between the two problems represented by the left and right sides of (9). Since the two sets of distributions are paired off exactly, they must have the same number, and hence (9) is a true equality.

In proving that the two sides of (9) are equal, we did not have to know the numerical value of either side. In this case, however, it is easy to list all possible distributions systematically and thus to find

$$\{3, 4 \mid 1, 1, 5\}_7 = \{1, 1, 5 \mid 3, 4\}_7 = 4.$$

This proof of (9) consisted merely in interchanging the names of the colors and the names of the urns. We do not need to go through the detailed proof of (8). The same interchange of colors and urns shows the duality of the two problems and therefore the general equality (8).

8. *The computation of the value of distributions in certain cases.* The computation of the value of $\{r, s, \ldots \mid a, b, \cdot : \cdot\}_n$ for arbitrary numbers r, s, $\cdots$, a, b, $\cdots$ is quite troublesome and we shall not attempt it. The symbols that arose in paragraphs 3 to 6 are not of the most general sort. They all have the peculiarity that either all the numbers in front of the vertical bar or all the numbers following the bar (or both) consist entirely of 1's. That is, either all the balls have different colors or each urn can hold just one ball. Because of the duality we need only consider the case where the balls all have different colors, and we may compute

$$\{1, 1, \cdots, 1 \mid a, b, c, \cdots\}_n.$$

The subscript n reminds us that there are n balls and that all the urns together will hold just n balls,

$$1 + 1 + \cdots + 1 = a + b + c + \cdots = n.$$

Since we are only considering symbols with 1's in front of the bar, we don't always need to write them in, and we can use the shorter notation,

$$\{1, 1, \cdots, 1 \mid a, b, c, \cdots\}_n = \{a, b, c, \cdots\}_n.$$

The numbers a, b, c, $\cdots$, representing the sizes of the urns, are subject to no restrictions other than that their total be n. Obviously we have

$$(11) \qquad \{n\}_n = 1$$

for there is only one way to put all the balls into a single urn. If we replace the single urn $\mathcal{N}$ of content n by two urns, $\mathcal{N}_1$ of content $n-1$ and $\mathcal{N}'$ of content 1, then one of the balls must be taken from $\mathcal{N}$ and put into $\mathcal{N}'$. The remaining $n-1$ balls are put into $\mathcal{N}_1$. The one ball for $\mathcal{N}'$ can be any one and, since they are all different, this gives us n possibilities. Therefore we have

$$\{n-1, 1\}_n = n.$$

In the same way we prove

$$(12) \qquad a\{a, b, c, \cdots\}_n = \{a-1, 1, b, c, \cdots\}_n.$$

Here the urns B, C, $\cdots$ of content b, c, $\cdots$ are left unchanged, but the urn A of content a is replaced by two urns, A_1 of content $a-1$ and A' of content 1. We can go from a distribution among A, B, C, $\cdots$ to a distribution among A_1, A', B, C, $\cdots$ by taking any one ball from A for A', putting the remaining $a-1$ balls of A into A_1, and leaving the balls in B, C, $\cdots$ unchanged. Again, since the balls are all different, there are a different possible choices for the ball that is to be put in A'. Therefore there are a times as many distributions among A_1, A', B, C, $\cdots$ as among A, B, C, $\cdots$ and that is exactly the meaning of (12).

Now we can replace A_1 by two urns, A_2 of content $a-2$ and A'' of content 1, and will clearly have

$$(a-1)\{a-1, 1, b, c, \cdots\}_n = \{a-2, 1, 1, b, c, \cdots\}_n.$$

In the same way we find

$$(a-2)\{a-2, 1, 1, b, c, \cdots\}_n = \{a-3, 1, 1, 1, b, c, \cdots\}_n$$

and a whole series of similar equations, of which the last is

$$2\{2, \underbrace{1, 1, \cdots, 1}_{a-2}, b, c, \cdots\}_n = \{\underbrace{1, 1, \cdots, 1}_{a}, b, c, \cdots\}_n.$$

In this last step we have broken a down into a series of 1's. If we multiply (12) and all the equations that follow it together then both sides have the common factors

$$\{a-1, 1, b, c, \cdots\}_n, \{a-2, 1, 1, b, c, \cdots\}_n, \cdots, \{2, \underbrace{1, 1, \cdots, 1}_{a-2}, b, c, \cdots\},$$

and these can all be cancelled out without altering the equality. In this way we find

$$a(a-1)(a-2) \cdots 2\{a, b, c, \cdots\}_n = \{\underbrace{1, 1, \cdots, 1}_{a}, b, c, \cdots\}_n.$$

In the product $a(a-1)(a-2) \cdots 2$ we shall reverse the order and use the notation [4]

$$a! = 1 \cdot 2 \cdot 3 \cdots (a-1) \cdot a.$$

Our result can then be written as

(13) $$a!\{a, b, c, \cdots\}_n = \{\underbrace{1, 1, \cdots, 1}_{a}, b, c, \cdots\}_n.$$

Exactly the same process can be used to replace the urn B by smaller ones. This gives

$$b!\{\underbrace{1, 1, \cdots, 1}_{a}, b, c, \cdots\}_n = \{\underbrace{1, 1, \cdots, 1}_{a}, \underbrace{1, \cdots, 1}_{b}, c, \cdots\}_n$$

which can be combined with (13) to yield

$$a!b!\{a, b, c, \cdots\}_n = \{\underbrace{1, 1, \cdots, 1}_{a+b}, c, \cdots\}_n.$$

By treating C the same way, and continuing through the rest of the urns, we finally find

(14) $$a!b!c! \cdots \{a, b, c, \cdots\}_n = \{1, 1, \cdots, 1\}_n,$$

where, obviously, the right hand symbol contains exactly n 1's.

If $a = n$ in (13) then there is only one urn, and (13) becomes

$$n!\{n\}_n = \{1, 1, \cdots, 1\}_n.$$

Therefore, using (11), we find

(15) $$n! = \{1, 1, \cdots, 1\}_n.$$

Using this value in (14), we then have the formula

(16) $$\{a, b, c, \cdots\}_n = \{1, 1, \cdots, 1 \mid a, b, c, \cdots\}_n = \frac{n!}{a!b!c! \cdots}$$

[4] $a!$ is read "a factorial".

for the computation of our symbol, at least in all the cases that arise from our examples.

9. We can now return to the special examples.

Example I. From (2) and (15) we have

$$P_n = \{1, 1, \cdots, 1 \mid 1, 1, \cdots, 1\}_n = \{1, 1, \cdots, 1\}_n = n!,$$

that is, n persons can be seated at n places in $n!$ different ways, or there are $n!$ permutations of n different things. The factorials that appear in these formulas increase very rapidly. The first 10 are:

$$
\begin{array}{ll}
1! = 1 & 6! = 720 \\
2! = 2 & 7! = 5{,}040 \\
3! = 6 & 8! = 40{,}320 \\
4! = 24 & 9! = 362{,}880 \\
5! = 120 & 10! = 3{,}628{,}800.
\end{array}
$$

Example II. From (3) and (16) we find

$$S = \frac{32!}{10! \, 10! \, 10! \, 2!}$$

for the number of different ways the cards can be distributed in a game of skat. This is indeed a large number. On computing, it turns out to be

$$S = 2{,}753{,}294{,}408{,}504{,}640.$$

The reader may be interested in making the similar computation for the hands of bridge.

Example III. Making use of the duality (8) as well as (5) and (16), we obtain

$$P^{(n)}_{a,b,c} = \frac{n!}{a! \, b! \, c!} \qquad (a + b + c = n).$$

If these are computed for $n = 4$, they can be used in (16) to give the expansion

$$
\begin{aligned}
(x + y + z)^4 = {} & x^4 + y^4 + z^4 \\
& + 4x^3y + 4x^3z + 4xy^3 + 4y^3z + 4xz^3 + 4yz^3 \\
& + 6x^2y^2 + 6x^2z^2 + 6y^2z^2 \\
& + 12x^2yz + 12xy^2z + 12xyz^2.
\end{aligned}
$$

Example IV. Finally (7) and (16) give the value

$$C^{(n)}_k = \frac{n!}{k!(n-k)!}$$

for the number of combinations of n things taken k at a time.

9. On Waring's Problem

The sequence of squares 1, 4, 9, 16, 25, $\cdots$ becomes less and less dense as we go further out. The gaps between the consecutive squares become longer and longer. Although many numbers are not squares, some of them can at least be considered as sums of *two* squares, for example, $13 = 9 + 4$, $41 = 25 + 16$, etc. But not every number can be written as a sum of two squares. If we try to express the number 6 as a sum of two squares, the available squares are 1 and 4, the only squares that are less than 6. Neither $1 + 1$ nor $4 + 4$ nor $1 + 4$ gives 6, so 6 requires at least three squares. In fact, 6 can be expressed as a sum of *three* squares, $6 = 4 + 1 + 1$. The same procedure shows that 7 cannot be expressed as a sum of three squares, since the smallest number that will suffice is *four* and $7 = 4 + 1 + 1 + 1$. For $8 = 4 + 4$, two squares again suffice, 9 is itself a square, and we have $10 = 9 + 1$, $11 = 9 + 1 + 1$, $12 = 9 + 1 + 1 + 1 = 4 + 4 + 4$, etc.

One would naturally expect that we would soon come to a point where *four* squares would no longer be enough, and that on continuing, more and more squares would be needed. However, Fermat, who ranks with Descartes among the greatest mathematicians of the 17th century, proved the very surprising fact that *every* positive whole number can be expressed as a sum of at most *four* squares.

Waring conjectured that a similar fact could be proved for cubes, fourth powers, etc. and he brought up the question as to how many cubes, fourth powers, and so on would be required. Because of this his name has been attached to this set of problems. The cubes are the numbers 1, 8, 27, 64, $\cdots$. If we try to express the smaller numbers as sums of cubes, we see that 7, being the last number before 8, must be expressed entirely by means of 1's. Therefore $7 = 1 + 1 + 1 + 1 + 1 + 1 + 1$ requires 7 cubes. Similarly, $15 = 8 + 1 + 1 + 1 + 1 + 1 + 1 + 1$ requires 8 cubes, and $23 = 8 + 8 + 1 + 1 + 1 + 1 + 1 + 1 + 1$ requires 9. Before we reach 31 a new cube, 27, becomes available, and the whole situation is changed. In fact, $31 = 27 + 1 + 1 + 1 + 1$ requires only 5 cubes.

C. G. J. Jacobi had the computer Dahse compile a list showing the decomposition of each number into a sum of the smallest possible number of cubes. This list revealed that after 23, the next number requiring 9 cubes is 239. These were the only numbers of the entire list that require 9 cubes, and the list extended to 12000.

The numbers requiring just 8 cubes were found to be 15, 22, 50, 114, 167, 175, 186, 212, 213, 238, 303, 364, 420, 428, 454, and then no more were found all the way up to 12000. More numbers were found to require 7 cubes, 7, 14, 21, 42, 47, 49, 61, 77, 85, 87, 103, · · ·, 5306, 5818, 8042, but even this series appears finally to be coming to an end. Continuations of this empirical work have only added further confirmation.

Such empirical work can prove nothing. It only serves to suggest that it is probably true that every number is a sum of at most 9 cubes, and that probably every number from a certain point on can be expressed as a sum of at most 8, or perhaps even 7, cubes. The latter statement, that 8 cubes suffice from some point on, was first proved by Landau by the use of difficult mathematical methods. After this was established, Wieferich proved the former statement.

Fourth powers appear to behave in the same way as cubes. ·The first few fourth powers are 1, 16, 81, 256, · · ·. Now 15 requires 15 fourth powers, 31 requires 16, 47 requires 17, 63 requires 18, and 79 requires 19. After this the new fourth power 81 intervenes and the picture is completely changed. The question would now be, do 19 fourth powers *always* suffice? Much work has been done on this problem. Liouville proved that 53 suffice and this number was slowly forced down to 47, 45, 41, 39, 38, and then Wieferich obtained 37. However, these were all far from the hoped-for 19.

The great German mathematician Hilbert attacked the general problem in a different way. He did not try to improve the previous results, but considered instead the whole set of problems connected with cubes, fourth powers, etc. He was able to prove at one stroke that not only for cubes and fourth powers, but for fifth, sixth, and all higher powers, there is a number that will suffice (like the 9 and 37 for cubes and fourth powers). Obviously this number is larger for higher powers.

Hardy and Littlewood in England used still different and highly complex methods to attack the problem. The fact that they showed that all numbers from a certain point on are sums of 19 fourth powers should give an idea of the power of their methods, and this is just one of their many far-reaching results. We have already seen that among the smaller numbers at least one requires 19 fourth powers. Hardy and Littlewood's result states that there is some number N such that all numbers from N on are certainly sums of at most 19 fourth powers, but this number N, as it is given by their proof, is so enormous that they did not bother to actually compute its value. In a sense this practically settles the case of fourth powers, since all one would have to do would be to test systematically all the numbers less than N to determine whether or not 19 fourth powers will suffice for *every* number. However, this number N is so tremendous that such a testing would far exceed the ability of any computer.

We have used considerable space in discussing the *facts* connected with this problem. This discussion should serve to give an idea of how empirical study can be used to help discover facts and to develop new theorems. We shall now attempt to give some idea of the *methods* connected with this problem, especially those which were used by Hilbert. Unfortunately the powerful methods of Hardy and Littlewood are much too advanced and complex to be included. Even the proofs that we shall mention are partially beyond the scope of this discussion, but it is possible to exhibit the ideas that are used.

As always, we start with a simple case. The equation $(a + b)(a - b) = a^2 - b^2$ is familiar from algebra. If one forgets it, it can easily be verified by actually multiplying out the left side. This equation is true no matter what two numbers a and b may be. An equation that is always true is called an "identity". A somewhat more complicated identity is:

$$(1) \qquad (a^2 + b^2)(c^2 + d^2) = (ac + bd)^2 + (ad - bc)^2.$$

In order to verify this we remember the formula

$$(x + y)^2 = x^2 + 2xy + y^2$$

and use it to multiply out the right hand side. We obtain

$$(a^2c^2 + 2acbd + b^2d^2) + (a^2d^2 - 2abdc + b^2c^2)$$
$$= a^2c^2 + a^2d^2 + b^2c^2 + b^2d^2 + 2abcd - 2abcd.$$

The last two terms cancel each other and the rest can be grouped to give $a^2(c^2 + d^2) + b^2(c^2 + d^2) = (a^2 + b^2)(c^2 + d^2)$, which is just the left side of (1).

This identity yields a result of some interest: if each of two numbers is a sum of two squares then their product is also a sum of two squares. For example, $13 = 9 + 4$ and $41 = 25 + 16$ are both of this form. Then, according to (1), we find

$$533 = 13 \cdot 41 = (3^2 + 2^2)(5^2 + 4^2) = (3 \cdot 5 + 2 \cdot 4)^2 + (3 \cdot 4 - 2 \cdot 5)^2 =$$
$$23^2 + 2^2,$$

and we have the product 533 expressed as a sum of two squares. Formula (1) can be used in the same way for any two numbers that are sums of two squares.

Euler, the great Swiss mathematician of the 18th century, discovered the following identity:

$$(a_1^2 + a_2^2 + a_3^2 + a_4^2)(b_1^2 + b_2^2 + b_3^2 + b_4^2)$$

(2)
$$= (-a_1b_1+a_2b_2+a_3b_3+a_4b_4)^2+(a_1b_2+a_2b_1+a_3b_4-a_4b_3)^2$$
$$+ (a_1b_3-a_2b_4+a_3b_1+a_4b_2)^2+(a_1b_4+a_2b_3-a_3b_2+a_4b_1)^2.$$

This identity can be verified without difficulty if both sides are multiplied out, making use of the familiar formula

$$(x_1+x_2+x_3+x_4)^2 = x_1^2+x_2^2+x_3^2+x_4^2+2x_1x_2+2x_1x_3$$
$$+2x_2x_3+2x_1x_4+2x_2x_4+2x_3x_4.$$

Formula (2) is similar to (1) in that it shows that if two numbers are each sums of four squares, their product is also a sum of four squares. Lagrange used this identity in a very beautiful proof of the theorem of Fermat that *every* positive whole number can be expressed as a sum of at most four squares. In the first place this remark shows that it is only necessary to show that every *prime* number is a sum of four squares, since it will then automatically follow for composite numbers. To discuss the prime numbers, Lagrange again made use of (2). For the moment we will accept the theorem of Fermat as established, and will make use of it. Lagrange's proof is given at the end of this chapter.

This theorem was used by J. Liouville in proving that every number is a sum of at most 53 fourth powers. He also made use of an identity:

$$6(x_1^2+x_2^2+x_3^2+x_4^2)^2$$
(3)
$$= (x_1+x_2)^4+(x_1+x_3)^4+(x_2+x_3)^4+(x_1+x_4)^4+(x_2+x_4)^4+(x_3+x_4)^4$$
$$+ (x_1-x_2)^4+(x_1-x_3)^4+(x_2-x_3)^4+(x_1-x_4)^4+(x_2-x_4)^4+(x_3-x_4)^4.$$

In order to verify this we first use the binomial theorem to expand

$$(x_1 + x_2)^4 = x_1^4 + 4x_1^3x_2 + 6x_1^2x_2^2 + 4x_1x_2^3 + x_2^4$$

and

$$(x_1 - x_2)^4 = x_1^4 - 4x_1^3x_2 + 6x_1^2x_2^2 - 4x_1x_2^3 + x_2^4.$$

Adding these, we have

$$(x_1 + x_2)^4 + (x_1 - x_2)^4 = 2x_1^4 + 2x_2^4 + 12x_1^2x_2^2.$$

Using the corresponding formula for each parenthesis in the second row of (3) plus the parenthesis below it, we find that the right side of (3) has the expansion

$$6(x_1^4+x_2^4+x_3^4+x_4^4)+12(x_1^2x_2^2+x_1^2x_3^2+x_2^2x_3^2+x_1^2x_4^2+x_2^2x_4^2+x_3^2x_4^2).$$

Now if the left side of (3) is also multiplied out it is immediately seen to have the same value.

Liouville made use of this identity in the following way. Let n be any positive whole number. It must be proved that n is a sum of at most 53 fourth powers. He begins by dividing n by 6 and finding the quotient and remainder, $n = 6x + y$ (if n is 29, the quotient x is 4, and the remainder y is 5). Here y will be one of the numbers 0, 1, 2, 3, 4, 5. At this point Liouville makes use of Fermat's theorem for the first time. He uses it to show that x can be written as a sum of four squares, $x = a^2 + b^2 + c^2 + d^2$. Then the original number can be written

$$n=6x+y=6(a^2+b^2+c^2+d^2)+y=6a^2+6b^2+6c^2+6d^2+y.$$

Now Liouville uses Fermat's theorem again, applying it to a, b, c, d to obtain

$$a = a_1^2 + a_2^2 + a_3^2 + a_4^2,$$
$$b = b_1^2 + b_2^2 + b_3^2 + b_4^2,$$
$$c = c_1^2 + c_2^2 + c_3^2 + c_4^2,$$
$$d = d_1^2 + d_2^2 + d_3^2 + d_4^2,$$

and therefore

$$n=6(a_1^2+a_2^2+a_3^2+a_4^2)^2 + \cdots + 6(d_1^2+d_2^2+d_3^2+d_4^2)^2 + y.$$

The identity (3) can now be used. Applied to the first expression on the right, it asserts that this expression can be written as a sum of 12 fourth powers. The same is true of each of the other three similar expressions. Thus far $4 \cdot 12 = 48$ fourth powers have been used and y must still be broken down into fourth powers. Since y is 0, 1, 2, 3, 4, or 5, it can be expressed as a sum of at most 5 fourth powers, each of which is 1. That gives a total of $48 + 5 = 53$ fourth powers.

Lagrange's Theorem states that every positive whole number N can be written as the sum of the squares of four whole numbers, $N = z_1^2 + z_2^2 + z_3^2 + z_4^2$. In proving this theorem, we shall make use of formula (2) of this chapter but will change the letters. If we take $a_1 = x_1$, $a_2 = x_2$, $a_3 = x_3$, $a_4 = x_4$, $b_1 = -y_1$, $b_2 = y_2$, $b_3 = y_3$, $b_4 = y_4$, we find

$$(x_1^2+x_2^2+x_3^2+x_4^2)\ (y_1^2+y_2^2+y_3^2+y_4^2)$$

$$(2') \quad =(x_1y_1 + x_2y_2 + x_3y_3 + x_4y_4)^2 + (x_1y_2 - x_2y_1 + x_3y_4 - x_4y_3)^2$$

$$+(x_1y_3 - x_3y_1 + x_4y_2 - x_2y_4)^2 + (x_1y_4 - x_4y_1 + x_2y_3 - x_3y_2)^2.$$

The proof can be broken up into a number of steps. We first prove:

Theorem 1. If A and B can each be written as the sum of four squares, then so can the product AB. This follows directly from (2), since if $A = x_1^2 + x_2^2 + x_3^2 + x_4^2$ and $B = y_1^2 + y_2^2 + y_3^2 + y_4^2$, then (2) shows that AB is a sum of the squares of four numbers, each of which is clearly a whole number. This result will allow us to concentrate our attention on the prime numbers, since it shows that we need consider only the factors of a number. We shall not try to prove the whole result for prime numbers at once, but shall start with:

Theorem 2. If p is a prime number greater than 2, then it is possible to find a whole number m such that $1 \leq m < p$ and such that mp can be written as a sum of four squares, $mp = x_1^2 + x_2^2 + x_3^2 + x_4^2$. This would be easy to prove if we did not insist on $1 \leq m < p$, since we could write $0 \cdot p = 0^2 + 0^2 + 0^2 + 0^2$ or $p \cdot p = p^2 + 0^2 + 0^2 + 0^2$. However, it will be important later to have $1 \leq m < p$.

To prove theorem 2, we first notice that p is an odd number, since it is prime and greater than 2. We write down the numbers 0^2, 1^2, 2^2, $\cdots$, $\left(\dfrac{p-1}{2}\right)^2$, divide each by p, and keep only the remainders. This gives us $\dfrac{p+1}{2}$ numbers r, each between 0 and $p - 1$. For $p = 11$ we would write down 0, 1, 4, 9, 16, 25. Dividing each by 11, we find the remainders r to be 0, 1, 4, 9, 5, 3. In every case all of these remainders will be different. If two were equal we would have two whole numbers, $x_1 > x_2$, between 0 and $\dfrac{p-1}{2}$, whose squares would yield the same remainder r when divided by p. That is, we would have $x_1^2 = q_1 p + r$ and $x_2^2 = q_2 p + r$. Subtracting, we find $x_1^2 - x_2^2 = (q_1 - q_2)p$ or $(x_1 - x_2)(x_1 + x_2) = (q_1 - q_2)p$. Since p is a prime, it must divide either $x_1 - x_2$ or $x_1 + x_2$, but this is impossible, since $x_1 - x_2$ and $x_1 + x_2$ are positive numbers less than p.

Now we take the remainders r, increase each by 1 and subtract from p. This again gives us $\dfrac{p+1}{2}$ numbers s between 0 and

$p - 1$, all of which are different. For $p = 11$ we would obtain 10, 9, 6, 1, 5, 7. At least one of the numbers s must equal a number of our previous set r, since r uses up $\dfrac{p + 1}{2}$ of the p numbers 0, 1, 2, $\cdots$, $p - 1$, leaving over only $\dfrac{p - 1}{2}$ numbers, while there are $\dfrac{p + 1}{2}$ numbers s. For $p = 11$ we find the numbers 1, 9, 5 in both r and s.

Let $R = S$ be equal numbers from r and s. R is the remainder on dividing some x^2, $0 \leq x \leq \dfrac{p - 1}{2}$, by p. S is obtained by dividing some number y^2, $0 \leq y \leq \dfrac{p - 1}{2}$, increasing the remainder by 1 and subtracting from p. That is, $x^2 = q_1 p + R$, $y^2 = q_2 p + r$, $S = p - (r + 1)$. Adding these three equations, we have $x^2 + y^2 + S = (q_1 + q_2 + 1)p + R - 1$. Since $R = S$, we can write this as $x^2 + y^2 + 1 = mp$ where $m = q_1 + q_2 + 1$. Also, since $0 \leq x \leq \dfrac{p - 1}{2}$ and $0 \leq y \leq \dfrac{p - 1}{2}$, we have $0 < mp \leq \left(\dfrac{p - 1}{2}\right)^2 + \left(\dfrac{p - 1}{2}\right)^2 + 1 = \dfrac{p^2 - 2p + 1}{2} + 1 = \dfrac{p^2 - 2p + 3}{2} < \dfrac{p^2}{2} < p^2$, and therefore $0 < m < p$. This proves theorem 2, since we have $1 \leq m < p$ and $mp = x^2 + y^2 + 1^2 + 0^2$.

For $p = 11$ we could take $R = S = 5$, finding $x = 4$, $y = 4$, $4^2 + 4^2 + 1^2 + 0^2 = 3 \cdot 11$. However, if we take $R = S = 1$, we find $x = 1$, $y = 3$, $1^2 + 3^2 + 1^2 + 0^2 = 1 \cdot 11$. In this case we have been able to write $p = 11$ as a sum of four squares. The method we have used may not always give us p as a sum of four squares, however, so we now prove:

Theorem 3. If p is a prime number greater than 2 and if m is the smallest positive whole number such that mp is a sum of four squares, then $m = 1$. We already have $m < p$ from 2. This smallest m cannot be even, for if it were we would have $x_1^2 + x_2^2 + x_3^2 + x_4^2 = mp$, an even number. Then either all the x's would be even, two would be even and two odd, or all would be odd. In the second possibility, we can suppose that we have numbered the x's so that x_1 and x_2 are even, while x_3 and x_4 are odd. Then in every case $(x_1 + x_2)$, $(x_1 - x_2)$, $(x_3 + x_4)$, and $(x_3 - x_4)$ are all even numbers. We then have

$$\left(\frac{x_1 + x_2}{2}\right)^2 + \left(\frac{x_1 - x_2}{2}\right)^2 + \left(\frac{x_3 + x_4}{2}\right)^2 + \left(\frac{x_3 - x_4}{2}\right)^2$$

$$= \frac{1}{2}(x_1^2 + x_2^2 + x_3^2 + x_4^2) = \frac{m}{2}p,$$

and all the numbers whose squares are taken on the left are whole numbers. That is, $\frac{m}{2}p$ can be written as a sum of four squares, and the even m was not the smallest possible value, as we had supposed.

We now know that the smallest m is odd and $m < p$. To prove $m = 1$, we suppose $m > 1$ and again show that it could be reduced. Since m is odd, we may suppose $m \geqq 3$. Then we have

$$(4) \qquad mp = x_1^2 + x_2^2 + x_3^2 + x_4^2.$$

We divide each x_k by m and obtain a remainder r_k, $0 \leqq r_k < m$. If $0 \leqq r_k \leqq \frac{m-1}{2}$, we write $y_k = r_k$. If $\frac{m+1}{2} \leqq r_k \leqq m - 1$, we write $y_k = r_k - m$. In both cases we have $x_k = q_k m + y_k$ and $-\frac{m-1}{2} \leqq y_k \leqq \frac{m-1}{2}$. Since $y_k = x_k - q_k m$, we also have, using (4),

$$\begin{aligned}
y_1^2 + y_2^2 + y_3^2 + y_4^2 &= x_1^2 + x_2^2 + x_3^2 + x_4^2 - 2m(x_1 q_1 + x_2 q_2 + x_3 q_3 + x_4 q_4) \\
(5) \quad &+ m^2(q_1^2 + q_2^2 + q_3^2 + q_4^2) = mp - 2m(x_1 q_1 + x_2 q_2 + x_3 q_3 + x_4 q_4) \\
&+ m^2(q_1^2 + q_2^2 + q_3^2 + q_4^2) = mn,
\end{aligned}$$

where n is a whole number. Furthermore, we have $n > 0$, since if $n = 0$ we would have $y_1 = y_2 = y_3 = y_4 = 0$. This would mean that each x is a multiple of m, and hence each x^2 is a multiple of m^2. From (4) we would then see that mp is a multiple of m^2, so p is a multiple of m. But this cannot be, since p is prime and $1 < m < p$. We also have $mn = y_1^2 + y_2^2 + y_3^2 + y_4^2 \leqq 4\left(\frac{m-1}{2}\right)^2 < m^2$, and hence $n < m$.

Multiplying (4) and (5), we find

$$(6) \qquad m^2 np = (x_1^2 + x_2^2 + x_3^2 + x_4^2)(y_1^2 + y_2^2 + y_3^2 + y_4^2)$$
$$= \text{the right side of } (2').$$

The first number whose square appears on the right side of (2') is

$$x_1 y_1 + x_2 y_2 + x_3 y_3 + x_4 y_4$$
$$= x_1(x_1 - q_1 m) + x_2(x_2 - q_2 m) + x_3(x_3 - q_3 m) + x_4(x_4 - q_4 m)$$
$$= x_1^2 + x_2^2 + x_3^2 + x_4^2 - m(x_1 q_1 + x_2 q_2 + x_3 q_3 + x_4 q_4)$$
$$= mp - m(x_1 q_1 + x_2 q_2 + x_3 q_3 + x_4 q_4)$$
$$= m z_1$$

where z_1 is a whole number. The second number whose square appears on the right side of $(2')$ is

$$x_1 y_2 - x_2 y_1 + x_3 y_4 - x_4 y_3$$
$$= x_1(x_2 - q_2 m) - x_2(x_1 - q_1 m) + x_3(x_4 - q_4 m) - x_4(x_3 - q_3 m)$$
$$= m(-x_1 q_2 + x_2 q_1 - x_3 q_4 + x_4 q_3)$$
$$= m z_2,$$

where z_2 is also a whole number. In a similar way, we find that the third and fourth numbers are $m z_3$ and $m z_4$. Putting these in (6), we have

$$m^2 n p = m^2 z_1^2 + m^2 z_2^2 + m^2 z_3^2 + m^2 z_4^2,$$
$$n p = z_1^2 + z_2^2 + z_3^2 + z_4^2.$$

Therefore np can be written as a sum of four squares, and we have already found $0 < n < m$. This shows that $m > 1$ was not the smallest possible value, as we had supposed. All that is left is $m = 1$, and theorem 3 is proved.

For the case $p = 11$, $x_1 = 4$, $x_2 = 4$, $x_3 = 1$, $x_4 = 0$, $m = 3$, we have $y_1 = 1$, $y_2 = 1$, $y_3 = 1$, $y_4 = 0$; $4 \cdot 1 + 4 \cdot 1 + 1 \cdot 1 + 0 \cdot 0 = 3z_1$, $z_1 = 3$; $4 \cdot 1 - 4 \cdot 1 + 1 \cdot 0 - 0 \cdot 1 = 3z_2$, $z_2 = 0$; $4 \cdot 1 - 1 \cdot 1 + 0 \cdot 1 - 4 \cdot 0 = 3z_3$, $z_3 = 1$; $4 \cdot 0 - 0 \cdot 1 + 4 \cdot 1 - 1 \cdot 1 = 3z_4$, $z_4 = 1$ and finally $3^2 + 0^2 + 1^2 + 1^2 = 1 \cdot 11$, $n = 1$.

Theorem 4. If p is a prime number it can be written as a sum of four squares. This is hardly more than a restatement of theorem 3. This is true if $p = 2$, since we have $2 = 1^2 + 1^2 + 0^2 + 0^2$. If $p > 2$ it is true by theorem 3.

We finally come to Lagrange's Theorem:

Theorem 5. Every whole number $n \geq 0$ can be written as a sum of four squares. This is true for $n = 0$ and $n = 1$, since we have $0 = 0^2 + 0^2 + 0^2 + 0^2$ and $1 = 1^2 + 0^2 + 0^2 + 0^2$. It is also true if n equals a prime number by theorem 4. All that are left are the composite numbers. If n is composite, we can factor it into a product of primes, $n = p_1 p_2 p_3 \cdots p_t$, where $p_1, p_2, p_3, \cdots, p_t$ are prime numbers, not necessarily distinct. Now, by theorem 4,

p_1 and p_2 can be written as sums of four squares. Then by theorem 1, the product p_1p_2 can also be written as the sum of four squares. Again by theorem 4, p_3 can be written as a sum of four squares and, by theorem 1, so can the product $p_1p_2p_3$. Continuing in this manner, we finally find that n can be written as a sum of four squares.

10.   On Closed Self-Intersecting Curves

1.   The curves that we shall discuss in this chapter are of a special kind. Although they may be quite complicated, they must satisfy certain conditions. First, they must be traversible in a single passage. That is, one should be able to draw the whole curve with a single stroke, starting at a given point and never taking the pencil from the paper until the curve is completely drawn. Second, they must be closed. That is, when one draws the curve he should be able to start at a given point, trace out the whole curve, and return to the original starting point just as the curve is completed. Finally, the curves may cross themselves any number of times, but when they do, they shall pass through such a crossing point only twice. The examples shown in Figs. 26 and 27 satisfy these conditions, that in Fig. 28 does not. The type of crossing point that is allowed

Fig. 26　　　　　　Fig. 27　　　　　　Fig. 28

(Figs. 26 and 27) is called a double point. The crossing of Fig. 28 is not allowed and is called a triple point [1]. In tracing out the

---

[1] Although the material presented here has many connections with the second chapter, the fundamental ideas are essentially different, and the reader will do well to consider the two chapters independently. In the second chapter a network of curves was given and ways to traverse it were discussed. Here the given curve is a "closed route" and there is no question of how it can be traversed. Also, in the words of the second chapter, the present curves have junctions of only the fourth order, now called double points. Finally, at a double point there is no choice as to how the parts of the curve can be connected; the four parts coming together at a double point must be connected in two pairs of opposite parts, since the curve is supposed to cross itself there.

complete curve, one clearly passes through each double point twice. If we designate each double point by a number then we can show the order in which we pass through them by means of a series of numbers. For example, Fig. 26 has the order 1 2 2 1, while Fig. 27 has 1 2 3 1 2 3. Since each double point is passed twice, each number must appear twice in the series. Gauss noticed that it is not true that just any series in which each number appears twice will represent the order of double points on a curve. In the case of two double points, we have encountered the ordering 1 2 2 1, but there is no curve with the order 1 2 1 2. This can be verified by trial and error for this very simple case.

The principal result of this section will be the theorem that *in the sequence each double point appears once in an even place, once in an odd place.* Expressed a little differently, this asserts that the two places where the double point appears are separated either by an even number of places or by none at all. This theorem immediately shows that 1 2 1 2 is impossible since there is just *one* place between the two 1's.

2. In order to prove the theorem, we consider an arbitrary double point Q of the curve A (Fig. 29). If we start from Q and follow the curve A, we will eventually return to Q. When we first return to Q we will have traversed a *part B* of the whole curve A.

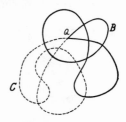

Fig. 29

This part can be *only a part* of the whole curve, since there are *four* segments of the curve radiating from Q and we have traversed only *two*, one when leaving Q and one when returning. If we continue from Q along the curve A we traverse the remainder, C, of A. Both B and C are closed curves, each with a sharp corner at Q. Although Q is a double point of A it is not a double point of B or C (neither curve crosses itself there), and B and C touch but do not cross each other at Q. We must prove that on traversing B from Q back to Q we pass through double points an even number of times.[2] The double points of A that are on B are, besides Q, the points where B crosses itself (double points of B) and the points where B and C intersect.

In traversing B one certainly passes through each double point of B twice. Therefore all these double points together contribute an even number to our count. At each intersection of B and C two paths cross, one a part of B, the other a part of C. In traversing B one never enters the path belonging to C, so one passes through each intersection of B and C only once. Now we need only show that B and C intersect in an even number of points.

3. Without changing the intersections of B and C we can slightly deform the curves near Q in such a way as to break them apart

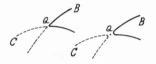

Fig. 30

(Fig. 30). The two curves now may cross each other at various points, but nowhere do they touch each other without crossing.

Now we have to show that two such curves either do not intersect, or intersect in an even number of points. In order to show this we shall deform one of the curves, say C, a step at a time. At each step we will remove a double point from C and thereby make the situation less complex. As at the point Q, we shall always deform the curve so little that we won't disturb the intersections of B and C.

Let P be a double point of C. The curve C splits into two closed curves D and E at P in the same way that A split into B and C at Q. The curves D and E touch each other at P. If we mark a certain direction on C, the direction along which we traverse it, then D and E are automatically given a direction (Fig. 31a). Now, starting at P, we traverse D in the marked direction, return to P, and then traverse E back to P in a direction *opposite* to that marked. By using the opposite direction on E we have eliminated the crossing at P.

---

[2] An even number of *times*, not of double points. In Fig. 26, with the order 1 2 2 1, we pass through the *single* point 2 between 1 and 1, but we pass through it *twice*.

We have traversed C in a single passage that went through P twice, along two paths which have corners at P but which do not cross

Fig. 31a　　　　　Fig. 31b　　　　Fig. 31c

there. These two paths merely touch each other at P so we can pull them slightly apart and round off the corners. We have then replaced C by a curve that does not have a double point at P (Fig. 31b). The new curve has one less double point than the original curve C. When deforming the curve near P, we must do it so slightly that no other double point of C and no intersection of B and C are disturbed.

We repeat this whole procedure for each double point of C and finally arrive at a curve C^* that is free of double points (Fig. 31c). The two curves C and C^* are nearly the same. They differ only near the double points of C. Of importance to us is the fact that both curves intersect B in the same points.

4. A closed curve that is free of double points encloses a region that we call the "interior" of the curve. This fact appears rather obvious, and we shall accept it on the basis of intuition. The part of the plane that is neither on the curve nor the interior is the "exterior," and it is separated from the interior by the curve itself. If such a curve is cut by another curve at some point, then the second curve must pass, at that point, from the interior to the exterior, or vice versa.

We can now show that the curves B and C either do not intersect, or intersect in an even number of points. Without altering the intersections, we can replace C by C^*, which has no double points. It may happen that B is entirely in the interior of C^*, in which case there are no intersections. If B lies entirely in the exterior of C^*, then again there are no intersections. Finally, B may lie partly in the interior of C^*, partly in the exterior. In that case let T be any point on B in the exterior of C^*. If we start at T and traverse B, there must be a first time that we enter the interior of C^*. This is one point of intersection of B and C^*. Since B is closed, we must finally arrive back at T in the exterior of C^*; that is, we must go from the interior to the exterior, a second intersection. It may happen that B enters the interior several times, but each time B enters the interior must be followed by a time that B leaves the interior, since B finally returns to the starting point T in the exterior. In all cases therefore, C^* and B (and hence C and B) either intersect in an even number of points or they do not intersect at all. As we saw at the end of § 2, this is enough to complete the proof of the theorem.

5. This theorem, which asserts that each double point of A occurs once at an even place and once at an odd place, can be expressed in a somewhat different way. We will think of A as the projection of a curve that is drawn in space, with the double points of A representing places where one part of the curve passes over or under another part of the curve. If the curve were a road the double points would represent underpasses. Now we would like to arrange the curve so that on traversing it we alternately take the upper and lower roads of the underpasses. If we start on the upper road of some particular underpass and traverse the curve, then we should go under the first underpass, over the next, etc. Thus the whole arrangement is completely determined. The only question is whether we can completely arrange it that way. For at some time we shall return to an underpass (a double point of A) that we have already crossed. Might it happen that we had originally crossed, say, over at this underpass and that we should again cross over in order to alternate? No, for our theorem states exactly that such a contradiction cannot occur. According to it we must have gone through double points of A, underpasses, an even number of times before returning to the original underpass. Since we started over the original underpass we went under the first, over the second, $\cdots$, over the last (because of the even number of times). Therefore on returning to the original underpass we should go under, and this is just the way which is left open.

Figs. 32 and 33 show the curves of Figs. 26 and 27 drawn in this way as projections of curves in space. At each double point they show which part of the curve passes over the other part. These space curves are "knots." A knot of this kind, in which we alternately go over and under when traversing its projection, is called an "alternating" knot. Strictly speaking, Fig. 32 is not a proper knot, for a loop of string twisted in that shape could be pulled out into an unknotted circle. However, Fig. 33 represents a proper knot. It cannot be changed to a circle without cutting it.

Fig. 34 shows that not all knots are alternating knots, at least without being deformed in some manner. The fact that there are

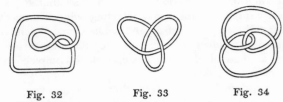

Fig. 32   Fig. 33   Fig. 34

non-alternating knots perhaps most clearly shows that our theorem, in either of its formulations, is certainly not trivial.

11. Is the Factorization of a Number into Prime Factors Unique?

1. Starting with any given number, one can keep splitting it into factors until one finally has only prime factors. For example, 60 may be factored as $6 \cdot 10$, 6 as $2 \cdot 3$, and 10 as $2 \cdot 5$, so that we finally have

$$60 = 2 \cdot 3 \cdot 2 \cdot 5$$

and these four factors are all prime.

Still using the example 60, we could first have factored it as $60 = 4 \cdot 15$, $4 = 2 \cdot 2$, $15 = 3 \cdot 5$, from which we have

$$60 = 2 \cdot 2 \cdot 3 \cdot 5.$$

The same primes appear in both these factorings and each prime appears the same number of times. Writing the primes in order of size, we have

$$60 = 2^2 \cdot 3 \cdot 5$$

in both cases. The fact that we get the same result in both cases seems very obvious because we are so used to it. In arithmetic we have learned to take it for granted that if we factor a number as far as possible into prime factors, we will always obtain the same factors no matter how we factor it.

That statement is true, but is it really so obvious? Consider a large number. It would take considerable work to factor 30031. After many trials we might discover that it can be factored into $59 \cdot 509$, and that these factors are prime. Now on what basis can we honestly say that it is obvious that further trials will not reveal some other factorization which is entirely different?

Such a question goes counter to all the ideas we have learned to accept concerning the prime numbers that "go into" a number. The aim of § 2 and § 3 will be to show that these intuitive ideas have no real basis. Then, having shown that there really is a problem, we shall devote the remainder of the chapter to proving that *the factorization into prime factors is indeed unique.*

2. In order to free ourselves of preconceived ideas we shall consider an unfamiliar system of numbers. These are the numbers of the form $a + b\sqrt{6}$, where a and b are ordinary whole numbers. For example, $12 + 5\sqrt{6}$, $\sqrt{6} - 2$, and $3\sqrt{6}$ are such numbers, while $2 + 3\sqrt{12}$ is not. The ordinary whole numbers are not excluded; in fact they merely have $b = 0$. They form a part of our number system, so this new system is an extension of the set of ordinary whole numbers.

Computation with these numbers is carried on just as one would naturally expect. The processes are all familiar from algebra. The way to add or subtract two of these numbers is made clear by the example

$$(3 + \sqrt{6}) + (5 + \sqrt{6}) = 8 + 2\sqrt{6}.$$

The following multiplications, carried out by the usual rules of algebra, should show how to multiply àny two numbers:

$$(3 + \sqrt{6})(3 - \sqrt{6}) = 9 - 6 = 3$$
$$(\sqrt{6} + 2)(\sqrt{6} - 2) = 6 - 4 = 2$$
$$(3 + \sqrt{6})(\sqrt{6} - 2) = 3\sqrt{6} - 6 + 6 - 2\sqrt{6} = \sqrt{6}$$
$$(3 - \sqrt{6})(\sqrt{6} + 2) = \sqrt{6}$$
$$(3 + \sqrt{6})(2 + \sqrt{6}) = 12 + 5\sqrt{6}$$
$$(3 - \sqrt{6})(\sqrt{6} - 2) = -12 + 5\sqrt{6}$$

We do not need to say anything about division. Sometimes one number will divide another, sometimes not, just as in the case of ordinary whole numbers.

In our number system, 6 can be factored into $\sqrt{6}\sqrt{6}$ as well as in the usual way,

$$(1) \qquad 6 = 2 \cdot 3 = \sqrt{6}\sqrt{6}.$$

Here, apparently, 6 can be factored in two different ways. This leads us to remember the question as to whether 30031 has a factorization that differs from $59 \cdot 509$. Apparently (1) presents an analogous situation.

However, this case can be cleared up in a very natural way. The numbers 2 and 3 are ordinary primes. They cannot be factored in the ordinary number system, but they can be factored in the new system. In fact, from our multiplication examples we have

$$2 = (\sqrt{6} + 2)(\sqrt{6} - 2), \qquad 3 = (3 + \sqrt{6})(3 - \sqrt{6}).$$

Therefore, continuing with our factorization of $6 = 2 \cdot 3$, we have

$$(2) \qquad 6 = (\sqrt{6} + 2)(\sqrt{6} - 2)(3 + \sqrt{6})(3 - \sqrt{6}).$$

The two factorizations (1) are merely (2) with different pairs of factors combined. The first of these has the first two factors and the last two combined. The second factorization has the first and fourth factors combined as well as the second and third. Obviously there are other possible combinations. For example, if we combine the first and third factors and the second and fourth factors, we have

$$6 = (12 + 5\sqrt{6})(-12 + 5\sqrt{6}),$$

and this factorization can be verified by direct multiplication.

This case is not different from what we are accustomed to. In clearing it up we did not have to know that the factors in (2) are prime. By a prime number, we now obviously mean a number that cannot be factored in our system. Actually it would not be difficult to show that the four factors are prime.

3. Now we turn to still another number system. This system is the set of numbers of the form $a + b\sqrt{-6}$, where a and b again are ordinary whole numbers. In this case we shall again find a situation like (1), but this time we will not be able to explain it away as we did with (2). Computations can be made in this system just as easily as in the system $a + b\sqrt{6}$. The rules of algebra are used exactly as they were before.

Corresponding to (1) we now have

$$(3) \qquad 6 = 2 \cdot 3 = -\sqrt{-6}\sqrt{-6}.$$

In analogy with the other case, we shall attempt to factor 2, 3, and $\sqrt{-6}$. This time however it will turn out that they are primes in this system, that we cannot factor them.

In our discussion it will be convenient to use the idea of the "norm" of a number. The norm of the number $a + b\sqrt{-6}$ is the product of that number with $a - b\sqrt{-6}$,

$$\mathcal{N}(a+b\sqrt{-6}) = (a+b\sqrt{-6})(a-b\sqrt{-6}) = a^2 + 6b^2.$$

In other words, in our system, the norm of a number is the product of that number and the number that is obtained by replacing $\sqrt{-6}$ by $-\sqrt{-6}$. The norm of a number is always a positive, ordinary whole number. Also the norm of the product of two numbers is equal to the product of their norms. For, according to the rule, we have

$$\mathcal{N}(a + b\sqrt{-6})(c + d\sqrt{-6})$$
$$= [(a+b\sqrt{-6})(c+d\sqrt{-6})] [(a-b\sqrt{-6})(c-d\sqrt{-6})],$$

and these four factors can be reordered to give

$$(a + b\sqrt{-6})(a - b\sqrt{-6})(c + d\sqrt{-6})(c - d\sqrt{-6}).$$

Pairing the terms, this is exactly

$$\mathcal{N}(a + b\sqrt{-6})\mathcal{N}(c + d\sqrt{-6}),$$

according to the rule.

If 2 could be factored into two factors in our system, we would have

$$2 = (a + b\sqrt{-6})(c + d\sqrt{-6}),$$

and therefore

$$\mathcal{N}(2) = \mathcal{N}(a + b\sqrt{-6})\mathcal{N}(c + d\sqrt{-6}).$$

But the norm of 2 is $\mathcal{N}(2) = (2 + 0\sqrt{-6})(2 - 0\sqrt{-6}) = 2 \cdot 2 = 4$, so we would have

$$4 = (a^2 + 6b^2)(c^2 + 6d^2).$$

That is, 4 would be factored into a product of two ordinary whole numbers, each of the form $x^2 + 6y^2$. There are only two ways in which 4 can be factored using ordinary numbers. Either both factors are 2, or one is 4 and the other 1. Neither helps us here, since 2 cannot be expressed in the form $x^2 + 6y^2$ and 1 is of this form only with $x = 1, y = 0$. Therefore the only way that 2 can be split into two factors in the system is for one factor to be $1 + 0\sqrt{-6} = 1$. We don't consider this as a factorization in this system any more than we would consider $5 = 1 \cdot 5$ as a factorization in the ordinary number system.

In exactly the same way one can recognize that 3 and $\sqrt{-6}$ are primes in the system. Instead of the norm 4, the norms 9 and 6 would have to be split into factors of the form $x^2 + 6y^2$.

We have now proved that (3) represents two different factorizations into prime factors of the number 6 in our system. If such a thing can occur in this system, then it is certainly not obvious that it cannot occur in the ordinary system. If there were any basis for asserting that it is obvious that factorization is unique in the ordinary number system, then, on the same basis, we could assert that it is obvious in every system. But we have found a system in which, far from being obvious, it is not even true. We shall see that factorization is unique in the ordinary system, but that it is a particular property of that system. In proving this, we will have to use particular properties of the system.

It is noteworthy that the Greek mathematicians recognized this problem and felt the need for proving it for the sake of logical completeness and clarity, apparently without the aid of an example such as ours. The unique factorization theorem is proved in Euclid but it is stated somewhat differently, without the use of modern notation. Beside the difference in formulation of the theorem, the proof given by Euclid is different from the one we shall use.

4. The number 30 is a multiple of 3. It is also a multiple of 5. This fact is expressed by saying that 30 is a "common multiple" of 3 and 5. In general, a common multiple of two numbers is a number that is simultaneously a multiple of each of the numbers. No matter what the two numbers are, they certainly have at least one common multiple, for if the numbers are a and b their product ab must be a common multiple. For 3 and 5, the product $3 \cdot 5 = 15$ is a common multiple, as is 30. The number 30 can be recognized as a common multiple of 3 and 5 by the fact that it is twice their product, 15. In the same way it is clear that every multiple of ab is a common multiple of a and b. Therefore two numbers always have an infinite number of common multiples.

The product ab and all its multiples together do not necessarily give us all the common multiples of a and b. For example, the multiples of $10 \cdot 15$ are 150, 300, 450, $\cdots$, while the common multiples of 10 and 15 are 30, 60, 90, 120, 150, 180, $\cdots$. Clearly 30 is the smallest number that is a common multiple of 10 and 15. There is always a smallest common multiple of any two numbers a and b, for one needs only to test each number from 1 to ab to decide which are common multiples. There will always be at least one common multiple, since ab itself is one. Among these common multiples there will be a smallest one, called the *least common multiple of a and b*.

We first prove

Lemma 1. Every common multiple of two numbers is a multiple of their least common multiple. For $a = 10$, $b = 15$ this asserts that the multiples of 30, that is 30, 60, 90, $\cdots$, are all the common multiples of 10 and 15. This can easily be verified in this case and in any other particular case. However, we must prove it *in general*.

The proof depends on the simple fact that the difference of two common multiples of a and b is again a common multiple of a and b. For the difference of two multiples of a is again a multiple of a; this was used and discussed in Chapter 1. The same is true with regard to b. Therefore, if each of two numbers is a multiple of both a and b, then their difference is also divisible by both and hence is a common multiple.

Let m be the least common multiple of a and b and let N be any common multiple. Then, by what we have just seen, $N - m$ is also a common multiple of a and b. If we again subtract m, we find that the same is true for $N - 2m$. Continuing to subtract m, we see that

$$N - m, \quad N - 2m, \quad N - 3m, \cdots$$

are all common multiples of a and b. Since m is the least of all the common multiples, the first number $N - m$ is certainly not negative. The same may be true for some of the following numbers, but eventually the numbers must be negative. Suppose that $N - xm$ is the last of these numbers that is positive. It is a common multiple of a and b and is not greater than m, since subtracting m gives the next number, which is no longer positive. Since m is the least common multiple, the only possibility is that $N - xm = m$. Therefore $N = xm + m = (x + 1)m$ is a multiple of m.

5. We can also speak of "common divisors" of two numbers. A number c is a common divisor of a and b if it divides both a and b exactly. According to lemma 1, the product ab, which is a common multiple of a and b, is a multiple of the least common multiple m. We can now prove

Lemma 2. The quotient of the product of two numbers a and b divided by their least common multiple, that is, the number

$$d = \frac{ab}{m},$$

is always a common divisor of a and b.

From the equation for *d* we have

$$d\,\frac{m}{a} = b$$

and m/a is actually a whole number since m is a multiple of a. Therefore b is a multiple of d or, in other words, d is a divisor of b. In exactly the same way, d is a divisor of a and hence a common divisor of a and b.

6. We can now prove a theorem from which we can immediately deduce the unique factorization theorem. We prove

Theorem: If a prime p divides the product xy of two numbers x and y, then p divides x or y, that is, it divides at least one of the factors.

We consider the least common multiple m of p and x. On the one hand the product xy is a multiple of p by hypothesis, and it is obviously a multiple of x. Therefore it is a common multiple of p and x and, by lemma 1, it is then a multiple of m,

(1) $$xy = hm.$$

On the other hand, by lemma 2, the number

(2) $$d = \frac{px}{m}$$

is a whole number and a common divisor of p and x. A divisor of a prime p can be only 1 or p. Therefore either $d = p$ or $d = 1$. In the first case $d = p$ is a divisor of x, so p divides the first factor x of xy. In the second case $d = 1$, and then (2) becomes $1 = \frac{px}{m}$ or $m = px$. Therefore, in virtue of (1), $xy = h(px)$. We can cancel the factor x to obtain $y = hp$. In this case p divides the second factor y of xy. In either case p divides at least one of the factors.

From this we have the

Corollary: If a prime divides a product of several numbers, then it divides at least one of the factors.

For if it divides, say, xyz, then it either divides x or yz. If it divides the latter then it divides either y or z. In any case, it divides one of the factors.

7. The unique factorization theorem follows at once. If we have two factorizations

$$N = pqrs \cdots = PQRS \cdots$$

of N into prime factors, then p divides N. Therefore p divides the right hand product. By the corollary it divides one of the prime factors. But if one prime divides another prime, the two must be equal because of the definition of a prime. Therefore p must occur somewhere among the primes on the right hand side. In the same way q and all the other primes on the left must appear on the right. Since the left and right sides are interchangeable, all the primes on the right must appear on the left. In other words, the two factorizations contain exactly the same primes.

Now we have only to see that each prime appears on both sides the same number of times. If p appears a times on the left and A times on the right we have the factorizations

$$N = p^a q^b r^c \cdots = p^A q^B r^C \cdots.$$

If a and A were different one of them (say A) would be larger. We could then divide by p^a to obtain

$$M = \frac{N}{p^a} = q^b r^c \cdots = p^{A-a} q^B r^C \cdots.$$

This would then represent two factorizations of M with p explicitly entering on the right but absent on the left. We have just shown that in two factorizations of any number the same primes must appear in both factorizations. In particular this must be true for M, so a and A could not have been different. Therefore we have $a = A$ and, in the same way, $b = B$, $c = C, \cdots$. That is, each prime appears in each factorization the same number of times.

Quite naturally one will wonder why this same proof does not hold in the number system $a + b\sqrt{-6}$ of § 3. In fact, nearly all of the proof can be carried over into that system. The one part that cannot be carried over is lemma 1. This lemma must therefore be the essential step in our proof.

12. The Four-Color Problem

1. In 1879 Cayley discussed the following problem. A map is usually printed in several colors in order to distinguish between the different countries. It would be best if each country were printed in a different color, but this is too costly. Instead it is customary

to use as few colors as possible, being careful that countries are always differently colored when they are next to each other. Fig. 35a represents the map of an island that requires three colors, blue for the sea and two colors for the two countries. Fig. 35b requires four colors. The three countries all touch the sea so none can be the same color as the sea. Since each country touches every other one, they require three different colors, making a total of four colors. Fig. 35c shows that four colors may be required even if we disregard the coloring of the sea. Here the inner country takes over the role of the sea in Fig. 35b. Fig. 36 again requires only

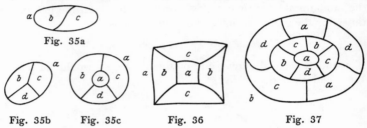

Fig. 35a

Fig. 35b   Fig. 35c   Fig. 36   Fig. 37

the three colors *a*, *b*, *c*, while the more complicated map of Fig. 37 can be colored with four colors. It would be natural to expect that more complicated maps would require more and more colors.

Many maps have been drawn, but regardless of how complicated they were, none that requires more than four colors has ever been found. On the other hand, no one has ever been able to prove that four colors will always be enough for every conceivable map. This is again a problem that can easily be proposed and understood without any mathematical preparation, but it is still unsolved.

However, it has been proved that *every map can be colored with five colors*. Whenever we say that a map can be colored we mean it in the sense that no two countries that have a boundary in common are to be given the same color. However, two countries may have the same color if they merely touch at a corner (as in the coloring of a checkerboard). Furthermore, by a country we mean a single piece of land and not a political subdivision made up of several separate parts.

The main purpose of this chapter will be to prove this fact, that every map can be colored with five colors. In the proof we shall assume that the map represents a single island. If each single island and the sea can be colored with five colors, certainly a map consisting of several islands can be colored by merely using exactly the same colors in each island.

2. As a first preliminary to the proof we shall prove *Euler's theorem*. This theorem is concerned with the number v of vertices (corners), f of faces (countries), and e of edges (boundaries) in an arbitrary map. It is a general theorem and has important applications other than the particular one at hand. This theorem, which was discovered by Euler but was already known to Descartes, asserts that

$$(1) \qquad v + f = e + 2.$$

In Fig. 36, for example, there are 8 vertices (i.e. points at which at least three countries come together), 6 faces, and 12 edges (each edge extends from one vertex to the next), and we have, in fact,

$$8 + 6 = 12 + 2.$$

In order to prove the theorem we shall discard for the moment the idea of a map and instead we shall think of the figure as representing a system of dikes and fields. The edges are now dikes separating the fields represented by the faces, and the outer area (originally the sea of the map) is covered with water. We now think of breaking down one dike after another until all the fields are under water (Fig. 38). In doing this it is not necessary to destroy *all* the dikes. Any dike that already has water on both sides of it can certainly be left. If we only break dikes that have water on just one side, then at each step we shall destroy one dike and flood one more field. The outer region (corresponding to the sea in the map) was under water at the start, so there are exactly $f - 1$ fields to be flooded. Since this process of flooding can certainly be carried out until all

Fig. 38

Fig. 39

Fig. 40

the fields are flooded, we shall finally have destroyed exactly $f - 1$ dikes.

We now wish to consider the system of dikes that has been left intact.

I. *One can walk dry-footed along the dikes from any vertex to any other vertex.* Before any dikes were destroyed this could certainly have been done, since we supposed at the start that the map represents a *single* island. Suppose that in the course of flooding the fields, the destruction of some dike *AB* (Fig. 39) would cut the system into two completely separated islands. If *AB* were destroyed it would be impossible to walk along dikes from *A* to *B*. That is, water would completely surround each of the two systems. Therefore there must be water on *both* sides of *AB* before it is destroyed, and it was explicitly stated that such a dike should not be destroyed.

II. *If one sends a messenger from any vertex P to any other vertex Q, then there is just one path available to the messenger.* For if there were two *different paths* from *P* to *Q*, then they would surround some area (Fig. 40). The ring of undestroyed dikes making up the two paths would keep this area dry, contrary to the fact that all the fields are flooded.

If we keep the starting point *P* fixed, there is just *one* path leading to each vertex. On each path there is a last edge that is passed over just before reaching the vertex. This edge is completely determined by the vertex. Therefore we have a correspondence between edges and vertices; corresponding to each edge is its end point. There are then just as many such end points as undestroyed dikes. The starting point *P* is not an end point, so the number of undestroyed dikes is $v - 1$. In all there are $f - 1$ destroyed dikes and $v - 1$ undestroyed. Therefore the total number e of dikes is

$$e = (f - 1) + (v - 1).$$

Euler's formula (1) follows at once if we remove the parentheses and transpose the numerical term.

3. As a final preliminary to the proof of the five-color problem we shall show that *it will be enough if we prove that five colors will color*

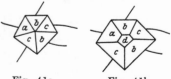

Fig. 41a   Fig. 41b

every map in which no more than three countries meet at a vertex. If we have a map in which more than three countries meet at some vertex (Fig. 41a), then we can draw another map (Fig. 41b) which is an exact copy of the original map except that one small new country has been formed around that vertex. This new map has one more country and several more vertices than the original, but only three countries meet at each new vertex and the vertex at which more than three countries meet has been eliminated. If we do the same thing for each vertex at which more than three countries meet, we will finally have a new map without any such vertices. Now if we can show that five colors are enough for every map in which no more than three countries meet at any vertex, five colors will be enough for the particular map we have just drawn. A possible coloring is shown by the letters in Fig. 41b. We can then color the original map by using the same color for each country as was used in the new map, disregarding the small countries that were added (Fig. 41a). It does not violate the rules if two countries that meet at a vertex but not along a boundary have the same color.

A vertex is a point at which *at least* three countries meet, but we have seen that we need consider only maps in which *no more* than three countries meet at a vertex. Therefore we need consider only maps in which *exactly* three countries meet at each vertex.

4. We are now ready to turn to the proof itself. We shall consider the number of vertices on the boundary of each country. If a given country has no vertices or just one vertex, then it has only one neighboring country and we can give it any color except that of its one neighbor. These countries can cause no difficulty, so we shall disregard them and assume that none are present in the rest of the proof.

Let f_2 be the number of countries with just 2 vertices, f_3 the number of countries with 3 vertices, etc. Then the total number f of all countries is the sum of the numbers of each kind,

$$(2) \qquad f = f_2 + f_3 + f_4 + \cdots.$$

The f_2 countries with 2 vertices each have 2 boundaries, that is, $2f_2$ boundaries altogether. The f_3 countries with 3 vertices each have 3 boundaries, $3f_3$ boundaries in all. And so on. This count will finally take into consideration *all* of the boundaries, but it will count each one twice, once for each of the two countries that are separated by it. Therefore we have

$$(3) \qquad 2e = 2f_2 + 3f_3 + 4f_4 + \cdots.$$

We can count the vertices in the same way. Since just three countries touch at each vertex we obtain

(4) $$3v = 2f_2 + 3f_3 + 4f_4 + \cdots$$

From (3) and (4) we have

(5) $$3v = 2e.$$

Euler's formula (1) multiplied by 6 is

$$6v + 6f = 6e + 12$$

and by (5), this is equal to

$$9v + 12.$$

Therefore we have

$$6f = 3v + 12$$

or, because of (2) and (4),

$$6(f_2 + f_3 + f_4 + \cdots) = (2f_2 + 3f_3 + 4f_4 + \cdots) + 12,$$

which simplifies to

(6) $$4f_2 + 3f_3 + 2f_4 + f_5 = 12 + f_7 + 2f_8 + \cdots.$$

From this we can show that *in every map in which just three countries meet at each vertex, there is a country having less than 6 vertices.* For if there were no such country, there would be no country with 2 vertices. Therefore $f_2 = 0$. In the same way, $f_3 = f_4 = f_5 = 0$. Therefore the left side of (6) would be 0, while the right hand side would be at least 12. The maps of Figs. 35a, 35b, 36, and 37 have no countries with more than 6 vertices, so the right side of (6) is 12 for all these cases. On the other hand, we have

in Fig. 35a:	$f_2 = 3,$	$f_3 = 0,$	$f_4 = 0,$	$f_5 = 0,$
in Fig. 35b:	$f_2 = 0,$	$f_3 = 4,$	$f_4 = 0,$	$f_5 = 0,$
in Fig. 36:	$f_2 = 0,$	$f_3 = 0,$	$f_4 = 6,$	$f_5 = 0,$
in Fig. 37:	$f_2 = 0,$	$f_3 = 0,$	$f_4 = 0,$	$f_5 = 12.$

In each of these examples there is only one of the numbers f_i that differs from 0. Fig. 35c is different since it has $f_2 = 0$, $f_3 = 2$, $f_4 = 3$, and all the rest are 0.

Now we know that in our map at least one country has less than 6 vertices. It may have 2, 3, 4, or 5 vertices, and we must take up each of these four possibilities in turn.

I. *There is a country with 2 vertices.* This country has just two neighbors. We can think of removing one of its two boundaries (the dotted one in Fig. 42). The new map has $f - 1$ instead of f countries. Let us suppose for the moment that this new map with fewer countries can be colored with only 5 colors. We can call a the color of the country formed by the original country and its neighbor when we removed the boundary. The color of the other neighbor can be called *b*. If we replace the boundary, then the original country with 2 vertices can be given the color *c*, since

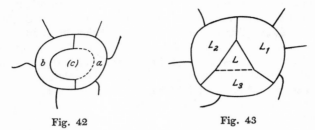

Fig. 42   Fig. 43

this country has only the two neighbours with colors *a* and *b*, leaving *c*, *d* and *e* available. Therefore, if the new map with $f - 1$ countries can be colored with 5 colors, then the original map with f countries can be colored with 5 colors as well.

II. *There is a country with 3 vertices.* Let *L* be the country and L_1, L_2, L_3 its three neighbors (Fig. 43). We remove one of the boundaries of *L* and suppose that the new map with $f - 1$ countries will require no more than 5 colors. Then, on replacing the boundary, we need only give *L* a color different from the colors of L_1, L_2, L_3. That leaves us 2 colors, either of which can be used for *L*.

III. *There is a country with 4 vertices.* Arguing in the same way, we see that here at least one of the 5 colors is left for *L*, since its 4 neighbors can use at most 4 different colors. However, a new kind of difficulty can arise in this case. A country such as L_2 of Fig. 44 might touch *L* along two *different* boundaries. If we remove one of

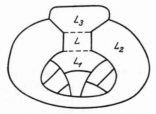

Fig. 44

these boundaries we must remove the other, for a boundary cannot separate a country from itself; that is not what we mean by a boundary. This would yield a country bounded by two completely separate boundaries and shaped like a ring. We have tacitly excluded such countries in our proof of Euler's theorem. However, we can avoid forming a ring-shaped country. If L and L_2 together form a ring, then the other two boundaries of L must belong to two countries L_1 and L_3 that are separated by the ring. Consequently L_1 and L_3 are different and have no boundary in common, so they can be given the same color. We remove the boundaries that separate L from L_1 and L_3 and obtain a new map with $f - 2$ countries. If this map with less than f countries can be colored with 5 colors, then so can the original, for L has only 3 neighbors and two of these have the same color. That leaves 3 colors available, any one of which can be given to L.

IV. *There is a country with 5 vertices.* The same difficulties can arise in an even more complicated form. Either one country may have two different boundaries in common with L (Fig. 45a), or two neighbors of L may touch each other along some distant boundary (L_2 and L_4 in Fig. 45b). In both cases L has two neighbors, L_1 and L_3, which are different and do not touch. This is clear, since they cannot cross the ring formed by L and L_2 in Fig. 45a

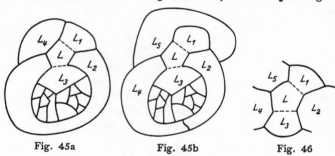

Fig. 45a   Fig. 45b   Fig. 46

or by L, L_2, L_4 in Fig. 45b. In both these cases as well as in the simple case where no ring occurs, L has two neighbors L_1 and L_3 which do not meet on a boundary (Fig. 46). Now we remove the boundaries between L and both L_1 and L_3. The new map has $f - 2$ countries, fewer than the original map. Let us again suppose that the new map can be colored with 5 colors and that $L + L_1 + L_3$ has the color a, L_2 has b, L_4 has c, L_5 has d. This is a total of 4 colors. On replacing the two boundaries, we can color L with the fifth color.

Every map will belong to one of these four cases, so our reduction process is complete. Every map can be reduced by removing one or two boundaries. The reduced map has fewer countries and, if it can be colored with 5 colors, so can the original map. We think of repeating the reduction and continuing until at most 5 countries are left. A map with no more than 5 countries can obviously be colored with 5 colors. The same is then true for each map of our reduction process, and hence is true for the original map.

5. We have proved the five-color theorem for maps drawn on a flat plane. However, the same proof can be used for maps on a globe. The outer country, that we usually thought of as the sea, fills the remainder of the globe. An ambiguity would occur only if we had not included the sea as one of the countries to be colored. Our proof remains the same step for step on the globe. We do not have to go over it all again to see this. The proof does not make any use of equality of line segments or angles. It does not involve any congruence theorem or any other idea that does not carry over directly to the surface of a sphere. The only concepts that are used are those connected with the relative positions of points, curves, and areas, and these are the same on a sphere as on a plane.

The situation is different if we consider a map on a surface shaped like a doughnut. On this surface it is possible to draw a map consisting of 7 countries, each of which has a boundary in common with all 6 other countries. Therefore this map requires 7 colors. It is not easy to visualize a map on this surface without having a model in one's hand, so we shall content ourselves with merely stating the facts. Certainly our proof of the five-color theorem cannot be carried over to this surface. Why does our proof easily carry over to the globe but suddenly fail for the doughnut?

There are two points at which our proof fails in the case of the doughnut, which is usually called a torus. The first is at the end of the proof of Euler's theorem where, with reference to Fig. 40, we said that two different paths from P to Q would surround an area. The second is in the proof in cases III and IV, where we said that L_1 and L_3 (Figs. 44, 45a, 45b) could not touch because they are separated by a ring made up of L and L_2 or L, L_2, and L_4. In both cases the proof on the plane or sphere depends on the fact that one cannot go from a point on one side of a closed curve to a point on the opposite side without crossing the curve. This is not true on the torus. In order to help us visualize the surface, let us think of the ring of Saturn as a solid rather than the loose mass of particles that it really is. This ring is a torus. Let us also suppose that a river flows around the ring, always remaining on the side that is away from Saturn (Fig. 47). If we are standing at A, on one bank of the river, can we walk to B on the opposite bank without crossing the river? All we need do is to walk directly away from the river

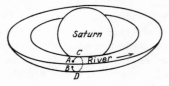

Fig. 47

to the point C, across the side nearest Saturn to D, and then under the ring and up to B.

This situation shows how careful we must be when relying on intuition and how easily our intuition can lead us astray. The part of the proof relating to Fig. 40 seemed intuitively obvious at the time, but now we see that it really requires a logical proof. This proof must depend on special properties of the plane and sphere, and it will certainly not remain valid for the torus.

We shall not go into this deeper analysis of the character of the plane and sphere. The properties that we have assumed can be proved, but it is not easy to do so. In the case of the torus it can be shown that Euler's formula must be replaced by

$$v + f = e.$$

Also, our proof of the five-color theorem can be carried over to the torus to show that 7 colors are enough to color every map on this surface. It is interesting to note that on the more complicated torus the color problem is completely solved, while on the simpler plane or sphere it is not known whether 5 or 4 colors are required.

13. The Regular Polyhedrons

1. We are going to make use of Euler's theorem to obtain an entirely different sort of result. We shall investigate the question: Do regular polyhedrons exist, and how many are there? A polyhedron is a solid figure bounded by portions of planes called faces. According to Euclid, a polyhedron is "regular" if all its faces are congruent polygons having equal sides and angles (regular polygons).

Our problem will be more extensive and the answer more satisfactory if we use a more general definition. We shall say that a polyhedron is "regular" if all its faces have the same number of sides, and if the same number of faces come together at each vertex (corner). In this definition we say nothing about equal sides, angles, or areas, nor anything about size. Only the *number* of certain parts is used in the definition.

We shall let φ represent the number of vertices possessed by each face. If the polyhedron is formed by triangles, then $\varphi = 3$, etc. The number of faces that meet at each vertex can be called ε. Since a polygon must have at least 3 vertices, we have

(1) $$\varphi \geqq 3.$$

In a solid figure a vertex is a point where at least 3 faces meet, so we also have

(2) $$\varepsilon \geqq 3.$$

We shall let v, f, and e represent the number of vertices, faces, and edges of the polyhedron respectively.

Let us imagine that the polyhedron is hollow and that it is made of some flexible substance such as rubber. We can then think of blowing the figure up until it becomes spherical. The faces of the polyhedron become pieces of the curved surface of the sphere and the edges become pieces of curved lines on the sphere. If we think of the sphere as a globe, then the original polyhedron has become a map representing f countries. Each country was formed from one of the faces. The boundaries of the map come from the edges of the poyhedron, so there must be e of them. Similarly, the map has the same number v of vertices as the original polyhedron. Each country has the same number φ of vertices and also of boundaries. Therefore only one of the numbers $f_2, f_3, \cdots$, (§ 4)*is different from 0, and this one must be equal to the total number of faces f. Exactly ε countries come together at each vertex. Examples of such maps are given in Figs. 35b, 36, 37. In these examples we have

$$\varphi = 3, \quad \varepsilon = 3; \qquad \varphi = 4, \quad \varepsilon = 3; \qquad \varphi = 5, \quad \varepsilon = 3$$

respectively.

By blowing up the polyhedron we have shown that Euler's theorem applies to polyhedrons as well as to maps. That is, *in every polyhedron we have the formula*

*Chap. 12

83

(3) $$v + f = e + 2$$

connecting the number of vertices, faces, and edges. Historically, it is in this form that Euler discovered the formula, and this is the form in which it is usually stated.

2. Each face of the regular polyhedron has φ vertices and φ edges. The f faces then account for $f\varphi$ edges in all, but here we count each twice, since an edge is a side of each of the two faces it separates. Therefore we have

(4) $$f\varphi = 2e.$$

Similarly, ε faces and hence ε edges meet at each vertex. That gives $v\varepsilon$ edges, but each is again counted twice because each edge has two ends. This gives us the formula

(5) $$v\varepsilon = 2e.$$

From (3) we have

$$v + f - e = 2,$$

and if it is multiplied by 2ε it becomes

$$2v\varepsilon + 2f\varepsilon - 2e\varepsilon = 4\varepsilon.$$

Now according to (4) we can replace $2e$ by $f\varphi$ and, since (4) and (5) show that $f\varphi = v\varepsilon$, we can also replace $v\varepsilon$ by $f\varphi$. This yields

$$2f\varphi + 2f\varepsilon - f\varphi\varepsilon = 4\varepsilon$$

or

(6) $$f(2\varphi + 2\varepsilon - \varphi\varepsilon) = 4\varepsilon.$$

Since f and 4ε are positive numbers, the same must be true of the factor in parentheses, so we have

(7a) $$2\varphi + 2\varepsilon - \varphi\varepsilon > 0.$$

Formulas (1) and (2) give *lower* bounds for φ and ε. Now we want to find upper bounds from (7a). First we can change the algebraic signs in (7a) to obtain

(7b) $$\varphi\varepsilon - 2\varphi - 2\varepsilon < 0.$$

If we compare the left side of (7b) with the product

(8) $$(\varphi - 2)(\varepsilon - 2) = \varphi\varepsilon - 2\varphi - 2\varepsilon + 4,$$

we see that it differs only by the added term 4. Consequently we shall add 4 to both sides of (7b):

$$\varphi\varepsilon - 2\varphi - 2\varepsilon + 4 < 4.$$

This, with (8), gives us

(9) $$(\varphi - 2)(\varepsilon - 2) < 4.$$

From (1) and (2) we see that the factors $(\varphi - 2)$ and $(\varepsilon - 2)$ of (9) are at least 1. Therefore $(\varphi - 2)$ and $(\varepsilon - 2)$ are positive whole numbers whose product is less than 4. There is no difficulty in finding all products of this kind. They are

(10) $$1 \cdot 1, \quad 1 \cdot 2, \quad 2 \cdot 1, \quad 1 \cdot 3, \quad 3 \cdot 1,$$

or five products in all. Every other product of two positive whole numbers would either be 4 or a larger number. *From the fact that there are only five products of two positive whole numbers which are less than four, we will deduce that there can be only five regular polyhedrons.*

3. In the five products (10) we can suppose that the first factor is $(\varphi - 2)$, the second $(\varepsilon - 2)$. We then have the five pairs of values for φ and ε:

φ	ε
3	3
3	4
4	3
3	5
5	3

We have obtained this table directly from Euler's theorem. Now if we remember the meaning of φ and ε, we see from the table that regular polyhedrons, if they exist, can have only triangles, quadrilaterals, or pentagons as faces. Furthermore, 3, 4, or 5 faces must meet at the vertices. Now from (6) we have

$$f = \frac{4\varepsilon}{2\varphi + 2\varepsilon - \varphi\varepsilon}$$

and, using (8), this can be written as

(11) $$f = \frac{4\varepsilon}{4 - (\varphi - 2)(\varepsilon - 2)}.$$

With (4) this gives

(12) $$e = \frac{f\varphi}{2} = \frac{2\varepsilon\varphi}{4 - (\varphi - 2)(\varepsilon - 2)}$$

and from (5) and (12) we find

(13) $$v = \frac{2e}{\varepsilon} = \frac{4\varphi}{4 - (\varphi - 2)(\varepsilon - 2)}.$$

The last three formulas give us just one value each for f, e, and v, corresponding to each of the five possible pairs of values of φ and ε. These values are listed in the following table:

φ	ε	f	e	v	
3	3	4	6	4	Tetrahedron
3	4	8	12	6	Octahedron
4	3	6	12	8	Hexahedron
3	5	20	30	12	Icosahedron
5	3	12	30	20	Dodecahedron

Therefore there are only five possible regular polyhedrons. They are also called the five Platonic solids and are named, as shown in the table, according to the number of faces.

4. This table has a very special property. If φ and ε are interchanged, the icosahedron and the dodecahedron are interchanged, as are the octahedron and hexahedron. The tetrahedron remains unchanged. At the same time e is unchanged but v and f are interchanged. We could have seen this from the previous formulas without reference to the table. Since the conditions (1), (2), and (9) are symmetrical in φ and ε, any admissible pair of values for φ and ε is still admissible if we interchange φ and ε. Furthermore, (12) is also symmetrical in φ and ε, so e is not altered by the interchange. Finally, (11) and (13) show that the interchange of φ and ε merely interchanges v and f.

This relationship is also clear for purely geometric reasons. We need only choose an arbitrary point inside each face of a polyhedron and take these points as vertices of a new polyhedron. The edges of the new polyhedron can be drawn between every two vertices that are on neighboring faces of the original polyhedron. Then exactly one vertex of the new polyhedron lies on each face of the original, exactly one new edge crosses over each old edge, and exactly one new face cuts off each old vertex. Therefore the number e is the same in both polyhedrons, while v and f are interchanged.

5. We have left one important point undiscussed. So far we have only seen that there can be *at most* five regular polyhedrons since, if all the faces have the same number of vertices and the same number of faces meet at each vertex, the number of faces, edges, and vertices must correspond to one of the five possibilities shown in the table.

While the table includes all possible regular polyhedrons, we have not yet determined whether these five polyhedrons can actually be constructed. It is quite conceivable that some further restrictions that we have not taken into consideration might exist. These further restrictions might eliminate one or more of the types listed in the table. In a word, we have discussed *necessary* but not *sufficient* conditions for our regular polyhedrons.

Actually, our discussion was concerned more with "regular" maps on a sphere than it was with the polyhedrons themselves. These maps are merely the polyhedrons after they have been blown up into spheres. Now we can actually exhibit a map corresponding to each type in the table. Fig. 35b represents the tetrahedron, Fig. 36 the hexahedron, and Fig. 37 the dodecahedron. Maps corresponding to the octahedron and icosahedron are shown in Fig. 48 and Fig. 49. Although all the maps are drawn on a flat

Fig. 48

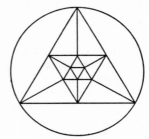

Fig. 49

plane, they may be transferred to a sphere. Now, if we disregard the deformation to which the polyhedrons have been subjected in blowing them up into spheres, we see that the conditions we have set up for regular polyhedrons are also sufficient.

This completes the discussion of our problem, but we have not shown that our figures can be constructed as regular polyhedrons according to the narrower definition of Euclid, that is, with congruent, regular plane polygons as faces. To prove this would require the use of concepts and theorems of an entirely different sort. Our investigation has used only such properties as remain unchanged under such deformations as stretching. Congruence is not such a property. It would require the use of metric geometry, in which equality of lengths and of angles is of importance, to show that our polyhedrons can be constructed according to Euclid's definition. We shall not carry out this proof. The existence of exactly five regular polyhedrons in the narrower sense dates from classical times. The proof is attributed to Theaetatus, a student of Plato's, and it is given by Euclid at the end of Book XII of his Elements.

14. Pythagorean Numbers and Fermat's Theorem

1. According to the Pythagorean theorem, the square on the hypotenuse of a right triangle has the same area as the sum of the squares on the two legs. Conversely, if three line segments are such that the square on one is equal to the sum of the squares on the other two, then the three segments will form a right triangle. The equation $a^2 + b^2 = c^2$ represents the fact that the segments of length a, b, c are the sides of a right triangle.

We have already seen in Chapter 4 that the hypotenuse and legs of an isosceles right triangle are incommensurable, that the equation $2a^2 = c^2$ can never be satisfied by whole numbers a and c. Are there any right triangles in which the sides are commensurable? In other words, can the equation

$$(1) \qquad a^2 + b^2 = c^2$$

be satisfied by three whole numbers? A simple and very well known example shows that the answer is yes:

$$3^2 + 4^2 = 5^2 \text{ or } 9 + 16 = 25.$$

Are there other answers? How can we find them? In this chapter we shall find the complete answer to these questions.

2. If we have a solution a, b, c of (1), we can easily find another by multiplying each of the terms a, b, and c by any whole number. Since 3, 4, 5 is a solution, we can multiply by 2 to find 6, 8, 10. This gives us

$$6^2 + 8^2 = 10^2.$$

More generally $3n$, $4n$, $5n$ will be a solution if n is any whole number. In the same way, if a, b, c is any solution, then an, bn, cn is also a solution, for from $a^2 + b^2 = c^2$ we have $a^2n^2 + b^2n^2 = c^2n^2$ or $(an)^2 + (bn)^2 = (cn)^2$. This way of finding new solutions is trivial and therefore not of much interest. It is more interesting to find the basic solutions, those that can't be found merely by multiplying another solution by some whole number. We shall call such solutions "reduced solutions", those solutions in which a, b, and c do not have a common divisor. Thus 3, 4, 5, is a reduced solution.

If two or more numbers have no common divisor we shall say that they are "relatively prime." In a reduced solution, each pair of the numbers a, b, c is relatively prime. For if a and b, say, had a common divisor d, they would also have every divisor of d as a common divisor. Some prime p would divide d; at worst, d would be p itself. Then a and b would have the common divisor p and we could write

$$a = pa_1, \quad b = pb_1.$$

Equation (1) would then become

$$p^2(a_1^2 + b_1^2) = c^2,$$

from which we see that p^2 would divide c^2. Then p would divide $c^2 = c \cdot c$, and hence one of the two equal factors c. That is, p would divide c as well as a and b. In the same way, a common prime factor of a and c or of b and c is a common factor of all three numbers.

3. Now we are looking for the solutions a, b, c of (1), in which every pair of a, b, and c is relatively prime. No two of the numbers can be even, that is, divisible by two; at most, only one can be even. Neither can all three numbers be odd, however. The square of an odd number $a = (2l + 1)$ is $a^2 = 4l^2 + 4l + 1$, which is again odd. Then, if a and b are odd, so are a^2 and b^2, so $a^2 + b^2$ is even and cannot be equal to the square of an odd c.

The only possibility that remains is that two of the numbers a, b, c are odd and one is even. Furthermore, we can see that c must be odd, for if c is even it is divisible by 2 and c^2 is divisible by 4. The other two numbers must be odd,

$$a = 2l + 1, \quad b = 2m + 1,$$

and we find

$$a^2 + b^2 = (4l^2 + 4l + 1) + (4m^2 + 4m + 1) = 4(l^2 + l + m^2 + m) + 2.$$

This number is even, but on division by 4, it leaves the remainder 2 and therefore could not equal c^2, which is divisible by 4.

We are now left with c odd and one of the numbers a and b even, the other odd. We shall let a be the odd number, b the even one. Thus in our example we have $a = 3$, $b = 4$, $c = 5$.

4. Equation (1) can be written in the form

$$(2) \qquad b^2 = c^2 - a^2 = (c + a)(c - a).$$

Here $(c + a)$ and $(c - a)$, being the sum and difference of two odd numbers, are both even. Their only common factor is 2. In other words, $\dfrac{c + a}{2}$ and $\dfrac{c - a}{2}$ are relatively prime, as we can see by supposing that d divides these two numbers. Then

$$\frac{c + a}{2} = df, \quad \frac{c - a}{2} = dg,$$

and on adding and subtracting these two equations we find

$$c = d(f + g), \quad a = d(f - g).$$

Then d divides both a and c, and this contradicts our assumption that they are relatively prime.

Since b, $c + a$, and $c - a$ are all even, we can write (2) in the form

$$(3) \qquad \left(\frac{b}{2}\right)^2 = \frac{c + a}{2} \cdot \frac{c - a}{2},$$

where the fractions are only apparent since each is actually a whole number. This equation expresses the square $\left(\dfrac{b}{2}\right)^2$ as a product of two *relatively prime* factors $\dfrac{c + a}{2}$ and $\dfrac{c - a}{2}$. We now come to the essential step of the proof. We prove that $\dfrac{c + a}{2}$ and $\dfrac{c - a}{2}$ must each be squares. If $\dfrac{b}{2}$ is factored into prime factors,

$$\frac{b}{2} = p^\alpha q^\beta r^\gamma \cdots,$$

where p, q, r, $\cdots$ are different primes, then we have

$$\left(\frac{b}{2}\right)^2 = p^{2\alpha} q^{2\beta} r^{2\gamma} \cdots.$$

All the prime factors of $\left(\dfrac{b}{2}\right)^2$ must appear in $\dfrac{c + a}{2}$ and $\dfrac{c - a}{2}$ taken together. However, each prime factor p must either appear only in $\dfrac{c + a}{2}$ or only in $\dfrac{c - a}{2}$, since $\dfrac{c + a}{2}$ and $\dfrac{c - a}{2}$ have no common factor. Therefore the prime factors of $\left(\dfrac{b}{2}\right)^2$ are distributed between

$\dfrac{c + a}{2}$ and $\dfrac{c - a}{2}$ in such a way that each prime power $p^{2\alpha}$, $q^{2\beta}$, $r^{2\gamma}$, $\cdots$

goes entirely into $\dfrac{c + a}{2}$ or $\dfrac{c - a}{2}$. Therefore $\dfrac{c + a}{2}$ and $\dfrac{c - a}{2}$ con-

tain only even powers of their prime factors and consequently each is a square.

5. We can now write

(4a)
$$\frac{c + a}{2} = u^2, \frac{c - a}{2} = v^2,$$

(4b)
$$\left(\frac{b}{2}\right)^2 = u^2 v^2,$$

where u and v, like u^2 and v^2, are relatively prime. From (4b) we have

(5)
$$b = 2uv,$$

and by addition and subtraction of the equations (4a) we find

(6)
$$c = u^2 + v^2, \ a = u^2 - v^2.$$

Since c and a are both odd, one of the squares u^2 and v^2 must be even and the other odd, in any other case their sum and difference would be even. The same must be true of u and v. We shall say that two such numbers are of "opposite parity".

We have now proved that if a, b, c is a reduced solution of (1), then a, b, and c can be represented in the form (5) and (6) by means of two numbers u and v which are relatively prime and of opposite parity. In our old example, $a = 3$, $b = 4$, $c = 5$, we have

$$u^2 = \frac{5 + 3}{2} = 4, \quad v^2 = \frac{5 - 3}{2} = 1,$$

$$u = 2, \quad v = 1,$$

and, in agreement with (5),

$$b = 4 = 2 \cdot 2 \cdot 1 = 2uv.$$

6. As yet we have only found *necessary* conditions for a reduced solution. We have started with a reduced solution a, b, c and have determined u and v. To complete the discussion we must show that the conditions are *sufficient*, that the a, b, c given by (5) and (6) is always a reduced solution when u and v are relatively prime and of opposite parity. We shall also add the condition $u > v$ to ensure that a is positive.

In the first place, a simple computation yields

$$(u^2 - v^2)^2 + (2uv)^2 = (u^2 + v^2)^2,$$

so the numbers given by (5) and (6) satisfy the equation (1). Secondly, to see that we have a *reduced* solution, we remember that u and v are relatively prime and one of them is even, the other odd. Equation (6) shows that a and c are both odd, so a, b, c do not have the common factor 2. They cannot have an odd common factor either, for if they did they would have, as common factor, an odd prime p (any prime other than 2). We could then write

$$c = pc_1, \quad a = pa_1,$$

and, from (4a), we would have

$$2u^2 = c + a = p(c_1 + a_1),$$
$$2v^2 = c - a = p(c_1 - a_1).$$

These equations imply that p divides both $2u^2$ and $2v^2$. Since p is different from 2 it would divide u^2 and v^2, but this is not compatible with the fact that u and v are relatively prime.

Some examples of Pythagorean numbers derived from (5) are given in the following list:

$$u = 2, \quad v = 1: \quad a = 3, \quad b = 4, \quad c = 5$$
$$u = 3, \quad v = 2: \quad a = 5, \quad b = 12, \quad c = 13$$
$$u = 4, \quad v = 1: \quad a = 15, \quad b = 8, \quad c = 17$$
$$u = 4, \quad v = 3: \quad a = 7, \quad b = 24, \quad c = 25$$
$$u = 5, \quad v = 2: \quad a = 21, \quad b = 20, \quad c = 29$$
$$u = 5, \quad v = 4: \quad a = 9, \quad b = 40, \quad c = 41$$

A glance at the table shows that not only is b even, but it is always a multiple of 4. This is true because $b = 2uv$, and either u or v is even.

7. Now that we have completely solved the problem connected with equation (1), a whole series of generalizations comes to mind. We can ask for a similar discussion of the equation

(7a) $$x^3 + y^3 = z^3,$$

or of

(7b) $$x^4 + y^4 = z^4,$$

or, more generally, of

(7c) $$x^n + y^n = z^n$$

for any $n > 2$. Pierre De Fermat (1601-1665) asserted that the equation (7c) has no solution in positive whole numbers x, y, z for $n > 2$. This statement, that has never been proved or disproved, is called Fermat's theorem, or Fermat's last theorem, to distinguish it from another Fermat theorem which we will mention in Chapter 23. However, the assertion has been proved for certain values of n. For example, it has been proved for all n from 3 to 100 by Kummer (1810-1893) and his followers. Before this, Euler (1707-1783) had proved it for (7a) and (7b).

With the aid of our knowledge of Pythagorean numbers, we can easily prove that (7*b*) cannot be solved in positive whole numbers. We shall even show that the equation

$$(8) \qquad x^4 + y^4 = w^2$$

cannot be solved in positive whole numbers. Since every fourth power is a square, but not every square is a fourth power, the insolubility of (8) is more significant than that of (7b).

In discussing the solutions, we insist on *positive* whole numbers in order to eliminate certain trivial solutions. For example, (8) is certainly satisfied by $x = 1$, $y = 0$, $w = 1$, while (7a) is satisfied by $x = -y$, $z = 0$.

8. If we write (8) in the form

$$(9) \qquad (x^2)^2 + (y^2)^2 = w^2,$$

we see that it is a special case of (1) with $a = x^2$, $b = y^2$, $c = w$. Since we need only consider "reduced" solutions x, y, z, we can see (as we did above) that if (9) has a solution w must be odd, while one of the squares x^2 and y^2 must be even and the other odd. We can let $x^2 = a$ be odd and $y^2 = b$ be even. If (9) has a solution, it must be given by (5) and (6). There must be two relatively prime numbers u and v, of opposite parity, determining the numbers

$$(10a) \quad x^2 = u^2 - v^2; \qquad (10b) \quad y^2 = 2uv; \qquad (10c) \quad w = u^2 + v^2.$$

Now (10a) can be written as

$$(11) \qquad x^2 + v^2 = u^2.$$

which is a new Pythagorean equation with x, v, u relatively prime. Here u takes the place of c, so it is odd. Since x is also odd, v must be even and it must take the place of b. The reduced solution x,

v, u of the Pythagorean equation (11) must again be given by (5) and (6) by means of new relatively prime numbers u_1 and v_1 having opposite parity. That is x, v, u are given by

$$(12) \qquad x = u_1^2 - v_1^2, \qquad v = 2u_1v_1, \qquad u = u_1^2 + v_1^2.$$

We now go back to equation (10b). Since u is odd, v even, and the two are relatively prime, we see that u and $2v$ are prime. This is because 2 does not divide u. Therefore (10b) expresses y^2 as a product of two relatively prime factors u and $2v$. According to the discussion in § 4, such a factorization of a square is possible only if each of the relatively prime factors is a square. Therefore we have

$$(13) \qquad u = w_1^2, \; 2v = 4t_1^2,$$

where we have made use of the fact that $2v$ is even. If we insert these two values in the last two equations of (12) we find

$$(14) \qquad t_1^2 = u_1v_1, \quad w_1^2 = u_1^2 + v_1^2.$$

Now u_1 and v_1 are relatively prime and their product is t_1^2, so they must again be squares,

$$(15) \qquad u_1 = x_1^2, \quad v_1 = y_1^2.$$

The second equation of (14) now becomes

$$(16) \qquad x_1^4 + y_1^4 = w_1^2.$$

9. Equation (16) is of the same type as the original equation (8). Starting with one reduced solution x, y, z (positive numbers) of (8), we have found another reduced solution for the same equation. The second solution is obtained from the first by means of a certain process. Without going through the whole process, we can see that the value of w in the first solution is larger than the value of w_1 in the second. For, by (10c) and the first equation of (13), we have

$$w = u^2 + v^2 = w_1^4 + v^2 > w_1^4$$

and hence $w > w_1$.

This will allow us to obtain a contradiction. Just as we went from x, y, w to x_1, y_1, w_1 in § 8, we can go from x_1, y_1, w_1 to another solution $x_2, y_{,2} w_2$, with $w_1 > w_2$ just as we had $w > w_1$. Repeating the process will give us yet another solution x_3, y_3, w_3 with $w_2 > w_3$.

Continuing in this way, we obtain a series of solutions with

(17) $$w > w_1 > w_2 > w_3 > \cdots .$$

These numbers are all positive whole numbers. There is only a finite number of positive whole numbers less than w, so the sequence (17) must finally end, say with w_k. But this w_k belongs to a solution x_k, y_k, w_k, and we can apply the process of § 8 to it to obtain still another solution $x_{k+1}, y_{k+1}, w_{k+1}$ with $w_k > w_{k+1}$. This contradicts the fact that (17) ends with w_k. Therefore we have established the fact that (8) does not have any solution in positive whole numbers, since the assumption of such a solution has led to a contradiction.

The basic idea of this proof was called the "principle of infinite descent" by Fermat. It consists of obtaining a contradiction by finding a process (in this case the repeated application of § 8) that yields a never-ending sequence of decreasing positive whole numbers. Such a sequence must end, since there is only a finite number of positive whole numbers less than n, the numbers $(n - 1), (n-2), \cdots, 3$ 2, 1, which are only $(n - 1)$ in number.

We have used the principle of infinite descent once before; it was essential to our proof of the irrationality of $\sqrt{2}$ in Chapter 4.

15. The Theorem of the Arithmetic and Geometric Means

A careful experimenter measures a certain object and finds its length to be 2.172 feet. On repeating his measurements twice more he obtains the lengths 2.176 ft. and 2.171 ft. What should he accept as the true length of the object? In such a case it is customary to use the average of the measurements, to add them and divide by their number. The experimenter would find the total to be 6.519 and, dividing by 3, would accept 2.173 ft. as the length of the object. This average that we have described is called the arithmetic mean. The arithmetic mean of n numbers $a_1, a_2, a_3, \cdots, a_n$ is

(1) $$A = \frac{1}{n} (a_1 + a_2 + a_3 + \cdots + a_n).$$

We shall soon see that there are other possible kinds of averages besides the arithmetic mean, but first let us see just what sort of thing an average should be. If the experimenter had made the same measurements in inches instead of feet, we would certainly agree that the average obtained should be the same as before, except that it would now be expressed in terms of inches. Each measurement would have been multiplied by 12 and the result of averaging should be multiplied by 12. The same thing should be true whether we multiply by 12 or by any other number t, since the experimenter could have used any sort of ruler to make his measurements. If the arithmetic mean (1) is to be a reasonable sort of average it must have this property: if each a is multiplied by t the final result A must also be multiplied by t. Thus,

$$\text{(A)} \qquad tA = \frac{1}{n} (ta_1 + ta_2 + ta_3 + \cdots + ta_n).$$

This is clearly true, since dividing both sides of (A) by t gives equation (1). We shall express this property (A) by saying that the arithmetic mean is "homogeneous". Anything that is to be used as an average should be homogeneous.

A second, simpler property that an average should have is this: the average of a_1, a_2, a_3, $\cdots$, a_n should neither be smaller than the smallest a nor larger than the largest a. It would be unreasonable for the experimenter to have obtained, say, 2.177 as the average of his measurements. The arithmetic mean (1) has this property, since the sum of n numbers is certainly no less than n times the smallest, and is no larger than n times the largest of the numbers.

There is one fairly obvious way in which the arithmetic mean can be changed. If the experimenter had made the last of his three measurements considerably more carefully than the other two, he might feel that he should give more weight to the last measurement, 2.171. If he felt that it deserved 4 times the weight of the other measurements he could list the first two measurements 2.172 and 2.176 and could then write down the third one 4 times. Now to average these he would add them all and divide by 6, obtaining 2.172 ft. It might have been easier to list the last measurement just once but to remember to multiply it by 4 and to divide the total by $1 + 1 + 4 = 6$. Such an average is called a weighted arithmetic mean. If a_1 is to be given a weight w_1, a_2 weight w_2, a_3 weight w_3, $\cdots$, a_n weight w_n then the weighted arithmetic mean is

$$W = \frac{1}{w_1 + w_2 + w_3 + \cdots + w_n} (w_1a_1 + w_2a_2 + w_3a_3 + \cdots + w_na_n).$$

It is quite easy to verify the fact that the weighted arithmetic mean has the two properties that should be shared by all averages. Although the topics which we shall discuss can be extended to weighted means, we shall not attempt this extension. It would only make the notation more cumbersome without adding anything to the essential ideas.

Turning to a new example, suppose we have 5 squares, two of side 1 in., one 2 in., one 5 in., and one 7 in. What is the average size of the squares? If we are interested in the *lengths of the sides* of the squares we merely find the arithmetic mean of the lengths, 3.2 in. If we are interested in the *areas* of the squares, however, we have two of area 1 sq. in., one of 4 sq. in., one of 25 sq. in., and one of 49 sq. in., and the arithmetic mean of the areas is 16 sq. in. This corresponds to a square of side 4 in. With respect to *sides* we would want to say that the average square has side 3.2 in., but with respect to *areas* the average square is larger, having sides of 4 in. If we think about how we obtained the number 4 we see that we squared the lengths, added these squares, divided by their number, 5, and took the square root of the result. Such an average is called the root mean square. The root mean square of the numbers a_1, a_2, a_3, $\cdots$, a_n is

(2) $$R = \sqrt{\frac{1}{n}(a_1^2 + a_2^2 + a_3^2 + \cdots + a_n^2)}.$$

This average also has the two properties that we demand of all averages. We shall only verify that it is homogeneous. If we multiply each a in (2) by t we have

$$\sqrt{\frac{1}{n}(t^2a_1^2 + t^2a_2^2 + t^2a_3^2 + \cdots + t^2a_n^2)} = \sqrt{\frac{1}{n}t^2(a_1^2 + a_2^2 + a_3^2 + \cdots + a_n^2)}$$

$$= t\sqrt{\frac{1}{n}(a_1^2 + a_2^2 + a_3^2 + \cdots + a_n^2)},$$

but this is just the R of (2) multiplied by t.

In the preceding example the arithmetic mean is not larger than the root mean square. This is not a mere accident of the example, but is true for any set of numbers a_1, a_2, a_3, $\cdots$, a_n. In order to prove this fact we shall first prove another fact, (6), which is of considerable interest in itself.

If any number, whether positive, negative, or zero, is squared, the result is either a positive number or zero. Therefore we have

$$0 \leq (c_1 - d_1)^2 = c_1^2 - 2c_1d_1 + d_1^2$$

no matter what the numbers c_1 and d_1 are. Adding $2c_1d_1$ to both sides of this inequality we have

$$2c_1d_1 \leq c_1^2 + d_1^2.$$

In a similar way we have

$$2c_2d_2 \leq c_2^2 + d_2^2,$$
$$2c_3d_3 \leq c_3^2 + d_3^2,$$
$$\cdots\cdots\cdots\cdots\cdots\cdots$$
$$2c_nd_n \leq c_n^2 + d_n^2.$$

Since the left side of each inequality is less than or equal to the right side, the sum of the left sides is less than or equal to the sum of the right sides,

$$2c_1d_1 + 2c_2d_2 + 2c_3d_3 + \cdots + 2c_nd_n$$
$$\leq c_1^2 + d_1^2 + c_2^2 + d_2^2 + c_3^2 + d_3^2 + \cdots + c_n^2 + d_n^2.$$

We can rewrite this in the form

$$(3) \quad 2(c_1d_1 + c_2d_2 + c_3d_3 + \cdots + c_nd_n)$$
$$\leq (c_1^2 + c_2^2 + c_3^2 + \cdots + c_n^2) + (d_1^2 + d_2^2 + d_3^2 + \cdots + d_n^2).$$

This holds for any numbers $c_1, c_2, c_3, \cdots, c_n, d_1, d_2, d_3, \cdots, d_n$. Now if $a_1, a_2, a_3, \cdots, a_n$ and $b_1, b_2, b_3, \cdots, b_n$ are also any numbers, the $a_1, a_2, a_3, \cdots a_n$ however not all zero, we can put

$$(4) \qquad r = \sqrt[4]{\frac{b_1^2 + b_2^2 + b_3^2 + \cdots + b_n^2}{a_1^2 + a_2^2 + a_3^2 + \cdots + a_n^2}},$$

$$c_1 = ra_1, \ c_2 = ra_2, \ c_3 = ra_3, \cdots, c_n = ra_n,$$

$$d_1 = \frac{b_1}{r}, \ d_2 = \frac{b_2}{r}, \ d_3 = \frac{b_3}{r}, \cdots, d_n = \frac{b_n}{r}.$$

Since $c_1d_1 = ra_1 \cdot \dfrac{b_1}{r} = a_1b_1, \ c_2d_2 = a_2b_2, \cdots, c_nd_n = a_nb_n$, (3) now becomes

$$(5) \quad 2(a_1b_1 + a_2b_2 + a_3b_3 + \cdots + a_nb_n)$$
$$\leq (r^2a_1^2 + r^2a_2^2 + r^2a_3^2 + \cdots + r^2a_n^2) + \left(\frac{b_1^2}{r^2} + \frac{b_2^2}{r^2} + \frac{b_3^2}{r^2} + \cdots + \frac{b_n^2}{r^2}\right).$$

Using (4), the first part of the right side can be written as

$$r^2(a_1^2 + a_2^2 + a_3^2 + \cdots + a_n^2)$$

$$= \sqrt{\frac{b_1^2 + b_2^2 + b_3^2 + \cdots + b_n^2}{a_1^2 + a_2^2 + a_3^2 + \cdots + a_n^2}} (a_1^2 + a_2^2 + a_3^2 + \cdots + a_n^2)$$

$$= \sqrt{(a_1^2 + a_2^2 + a_3^2 + \cdots + a_n^2)(b_1^2 + b_2^2 + b_3^2 + \cdots + b_n^2)}.$$

Similarly, the second part of the right side of (5) is

$$\frac{1}{r^2}(b_1^2 + b_2^2 + b_3^2 + \cdots + b_n^2)$$

$$= \sqrt{\frac{a_1^2 + a_2^2 + a_3^2 + \cdots + a_n^2}{b_1^2 + b_2^2 + b_3^2 + \cdots + b_n^2}} (b_1^2 + b_2^2 + b_3^2 + \cdots + b_n^2)$$

$$= \sqrt{(a_1^2 + a_2^2 + a_3^2 + \cdots + a_n^2)(b_1^2 + b_2^2 + b_3^2 + \cdots + b_n^2)},$$

and (5) becomes

$$2(a_1b_1 + a_2b_2 + a_3b_3 + \cdots + a_nb_n)$$

$$\leqq 2\sqrt{(a_1^2 + a_2^2 + a_3^2 + \cdots + a_n^2)(b_1^2 + b_2^2 + b_3^2 + \cdots + b_n^2)}.$$

Dividing each side by 2 we have

$$(6) \quad a_1b_1 + a_2b_2 + a_3b_3 + \cdots + a_nb_n$$

$$\leqq \sqrt{(a_1^2 + a_2^2 + a_3^2 + \cdots + a_n^2)(b_1^2 + b_2^2 + b_3^2 + \cdots + b_n^2)}.$$

This is Cauchy's inequality, named after the French mathematician A. L. Cauchy.

In order to show that the arithmetic mean is not larger than the root mean square, we remember that the a and b in (6) can be any numbers and we take $b_1 = b_2 = b_3 = \cdots = b_n = \dfrac{1}{n}$. We then have

$$\frac{a_1}{n} + \frac{a_2}{n} + \frac{a_3}{n} + \cdots + \frac{a_n}{n}$$

$$\leqq \sqrt{\left(a_1^2 + a_2^2 + a_3^2 + \cdots + a_n^2\right)\left(\frac{1}{n^2} + \frac{1}{n^2} + \frac{1}{n^2} + \cdots + \frac{1}{n^2}\right)}$$

where there are exactly n terms $\dfrac{1}{n^2}$ in the last parenthesis on the right. This is equal to $n \cdot \dfrac{1}{n^2} = \dfrac{1}{n}$ and our inequality becomes

$$\frac{1}{n}(a_1 + a_2 + a_3 + \cdots + a_n) \leqq \sqrt{\frac{1}{n}(a_1^2 + a_2^2 + a_3^2 + \cdots + a_n^2)}.$$

This is just what we wanted to prove, since the left side is the arithmetic mean (1) and the right side is the root mean square (2)

It is worthwhile to review the method by which we proved (6). The essential idea is that the square of a number is never negative. From this we obtained a number of inequalities whose sum is (3). Now (3) was not what we were looking for, but we were then able to obtain (6) merely by replacing the letters c and d with certain combinations of other letters. This idea of replacing one set of letters with another is often used in the study of inequalities, sometimes with very surprising results. In the proof of (6) it would have been possible, although not quite so simple, to prove (6) first and then to obtain (3) by changing the letters in (6).

In order to introduce one more sort of average we shall look at another example of measuring. An object is to be weighed by putting it on one pan of a balance and counterbalancing it with weights on the other pan. If the balance is not correctly built or has been damaged there may be a slight difference in the lengths of its two arms and this will make the weighing incorrect. It is usually impossible to measure the lengths of the arms accurately, so we shall make two weighings, one with the object in the left pan, one with it in the right pan. If the results of these two weighings are a_1 and a_2, what should we take as the true weight? Should we use the arithmetic mean, or is some other kind of average better? To answer this question let us suppose that the length of the left arm of the balance is l and that of the right arm is r. We cannot measure these lengths but the arms must have certain lengths. Now it is shown in elementary physics that the product of the weight by the length of the arm on one side is equal to the corresponding product on the other side. If w is the true weight of the object this means $wl = a_1r$ for the first weighing and $wr = a_2l$ for the second. If we multiply the left sides of these equations, and multiply the right sides we find $wlwr = a_1ra_2l$ or, dividing by lr, $w^2 = a_1a_2$. The true weight of the object is given by the formula $w = \sqrt{a_1a_2}$. This is a new sort of average. It is called the geometric mean of the numbers a_1 and a_2.

What is meant by the geometric mean of more than two numbers? If we are given n positive numbers $a_1, a_2, a_3, \cdots, a_n$, we will first multiply them together to obtain $a_1a_2a_3 \cdots a_n$. Should we then take the square root of this number? To decide this, we remember that an average must be homogeneous. If we multiply each a by t this product becomes $ta_1 \cdot ta_2 \cdot ta_3 \cdots ta_n = t^n a_1 a_2 a_3 \cdots a_n$. If the final result is merely to be multiplied by t, we must take the n-th root, not the square root. We can now say that the geometric mean of the n positive numbers a_1, a_2, a_3, $\cdots$, a_n is

$$(7) \qquad G = \sqrt[n]{a_1 a_2 a_3 \cdots a_n}.$$

One of the most famous theorems connected with means is the theorem of the geometric and arithmetic means. It states that the geometric mean of n positive numbers is never larger than the arithmetic mean, $G \leqq A$. There are several proofs of this theorem. We shall reproduce a particularly interesting and simple proof given by Cauchy.

We first prove the theorem for two numbers a_1 and a_2. We have

$$\left(\frac{a_1 + a_2}{2}\right)^2 = \frac{1}{4}\,(a_1^2 + 2a_1 a_2 + a_2^2) = \frac{1}{4}\,(a_1^2 - 2a_1 a_2 + a_2^2 + 4a_1 a_2)$$

$$= \frac{1}{4}\,(a_1^2 - 2a_1 a_2 + a_2^2) + a_1 a_2 = \left(\frac{a_1 - a_2}{2}\right)^2 + a_1 a_2.$$

Now since the square $\left(\dfrac{a_1 - a_2}{2}\right)^2$ is not negative, we see that $\left(\dfrac{a_1 + a_2}{2}\right)^2$ is $a_1 a_2$ increased by an amount that is not negative. That is,

$$(8) \qquad a_1 a_2 \leqq \left(\frac{a_1 + a_2}{2}\right)^2.$$

The square root of the left side is G and the square root of the right side is A, so we have proved the theorem for any two numbers, a_1 and a_2.

Now we want to prove the theorem for more numbers. It would be natural to try three numbers but as it turns out, it is easier to prove it for four. Since (8) is true for any numbers a_1 and a_2, we can replace a_1 by a_3 and a_2 by a_4 and have

$$a_3 a_4 \leqq \left(\frac{a_3 + a_4}{2}\right)^2.$$

From this and (8) we have

$$a_1 a_2 a_3 a_4 \leqq \left(\frac{a_1 + a_2}{2}\right)^2 \left(\frac{a_3 + a_4}{2}\right)^2,$$

which can be written in the form

$$(9) \qquad a_1 a_2 a_3 a_4 \leqq \left\{ \left(\frac{a_1 + a_2}{2}\right)\left(\frac{a_3 + a_4}{2}\right) \right\}^2.$$

Again, since (8) holds or any numbers, we can replace a_1 by $\left(\frac{a_1 + a_2}{2}\right)$ and a_2 by $\left(\frac{a_3 + a_4}{2}\right)$ to get

$$\left(\frac{a_1 + a_2}{2}\right)\left(\frac{a_3 + a_4}{2}\right) \leqq \left\{ \frac{\left(\frac{a_1 + a_2}{2}\right) + \left(\frac{a_3 + a_4}{2}\right)}{2} \right\}^2 = \left(\frac{a_1 + a_2 + a_3 + a_4}{4}\right)^2.$$

Using this in (9) we find

$$(10) \qquad a_1 a_2 a_3 a_4 \leqq \left(\frac{a_1 + a_2 + a_3 + a_4}{4}\right)^4.$$

This is our theorem for four numbers, since the 4-th root of the left side is G and the 4-th root of the right side is A.

We can repeat this argument again to prove the theorem for eight numbers. Using (10) we have

$$a_1 a_2 a_3 a_4 a_5 a_6 a_7 a_8 \leqq \left(\frac{a_1 + a_2 + a_3 + a_4}{4}\right)^4 \left(\frac{a_5 + a_6 + a_7 + a_8}{4}\right)^4,$$

and we can apply (8) to the right side as before to obtain

$$a_1 a_2 a_3 a_4 a_5 a_6 a_7 a_8 \leqq \left(\frac{a_1 + a_2 + a_3 + a_4 + a_5 + a_6 + a_7 + a_8}{8}\right)^8.$$

This gives us the theorem for eight numbers.

If we continue the proof in exactly the same manner, we see that $G \leqq A$ for the set of n positive numbers $a_1, a_2, a_3, \cdots, a_n$ if n is 2, 4, 8, 16, $\cdots = 2, 2^2, 2^3, 2^4, \cdots$, that is, if n is a power of 2. We must still prove the theorem for $n = 3, 5, 6, \cdots$, the numbers that are not powers of 2. This is now fairly easy. If $a_1, a_2, a_3, \cdots, a_n$ are any positive numbers, we can find a power 2^m of 2 which is larger than n (if $n = 50$ for example we take $m = 6$ since $2^6 = 64 > 50$). We then write

$$b_1 = a_1, \ b_2 = a_2, \ b_3 = a_3, \cdots, b_n = a_n,$$
$$b_{n+1} = b_{n+2} = \cdots = b_{2^m} = A,$$

where A is the arithmetic mean (1) of the a's. Now $b_1, b_2, b_3, \cdots, b_{2^m}$ are a set of 2^m positive numbers; so, from what we have already proved, their geometric mean is no larger than their arithmetic mean,

$$\sqrt[2^m]{b_1 b_2 b_3 \cdots b_{2^m}} \leqq \frac{1}{2^m} (b_1 + b_2 + b_3 + \cdots + b_{2^m}).$$

Raising both sides of this inequality to the 2^m-th power and inserting the values of the b's, we have

$$a_1 a_2 a_3 \cdots a_n A A \cdots A$$
$$\leqq \left\{ \frac{1}{2^m} (a_1 + a_2 + a_3 + \cdots + a_n + A + A + \cdots + A) \right\}^{2^m}$$

or

$$a_1 a_2 a_3 \cdots a_n A^{2^m - n} \leqq \left\{ \frac{1}{2^m} (a_1 + a_2 + a_3 + \cdots + a_n) + \frac{2^m - n}{2^m} A \right\}^{2^m}.$$

By (1) the right side reduces to

$$\left\{ \frac{1}{2^m} \cdot nA + \frac{2^m - n}{2^m} A \right\}^{2^m} = \left\{ \frac{n + 2^m - n}{2^m} A \right\}^{2^m} = A^{2^m},$$

and we can divide both sides by $A^{2^m - n}$ to obtain

$$a_1 a_2 a_3 \cdots a_n \leqq A^n.$$

Taking the n-th root of each side, we have

$$\sqrt[n]{a_1 a_2 a_3 \cdots a_n} \leqq A = \frac{1}{n} (a_1 + a_2 + a_3 + \cdots + a_n),$$

or $G \leqq A$, and this is the theorem of the arithmetic and geometric means.

16.   The Spanning Circle of a Finite Set of Points

1.   We consider a finite set consisting of n points $P_1, P_2, \cdots, P_n$, in a plane.   The distances between each pair of points P_i and P_j can be measured.   There must be a largest distance among this finite [1] set of distances.   This largest distance is called the "span" of the set of points.

[1] Using Chap. 8, § 9, IV, it is easily seen that one can find just $\dfrac{n(n-1)}{2}$ different pairs among the n points.

If a set of n points has span d, then we can draw a circle of radius d that completely surrounds the n points (Fig. 50). All we need

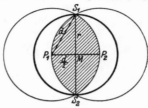

Fig. 50

do is draw a circle of radius d with any one of the n points, say P_1, as center. Since the distance of every other point from P_1 is at most d, the circle encloses all the other points P_2, P_3, $\cdots$, P_n, as well as its center P_1.

However, it is easy to construct a smaller circle that still encloses all the n points. First we find the pair of points whose distance apart is d. If there are several such pairs we can choose any one of them. Calling these two points P_1 and P_2, we draw two circles of radius d about them (Fig. 50). The circle with P_1 as center passes through P_2 and vice versa. Now all n points of the set lie in each circle, so they lie in the part of the plane that is common to both circles, the part that is shaded in the figure. If the two circles intersect at S_1 and S_2, then the circle having S_1S_2 as diameter encloses the whole common area and hence the whole set of n points. The radius r of this new circle can be found by applying the Pythagorean theorem to the triangle P_1MS_1

$$r^2 = d^2 - \left(\frac{d}{2}\right)^2 = \frac{3}{4}\, d^2, \; r = \frac{d}{2}\, \sqrt{3}.$$

We shall call a circle that surrounds all n points of the set an "enclosing" circle of the set. Besides the original enclosing circle of radius $r = d$, we have found an enclosing circle of radius $r = \frac{d}{2}\, \sqrt{3} = 0.866 \cdots d$.

2. Can this number $\frac{1}{2}\sqrt{3}$ be replaced by a still smaller one? The answer is given by the following theorem, discovered by H. W. E. Jung: Every finite set of points of span d has an enclosing circle of radius no greater than $\frac{d}{3}\, \sqrt{3} = 0.577 \cdots d$. There are many finite sets of points with smaller enclosing circles, but some require

a circle this large. On the other hand, since two points cannot be further apart than the diameter $2r$ of the enclosing circle, we have $2r \geqq d$, and therefore the enclosing circle can never have radius less than $\dfrac{d}{2}$. The proof of Jung's theorem will be the aim of this chapter.

An enclosing circle of radius $\dfrac{d}{3}\sqrt{3}$ can easily be found for the set of three points that form the vertices of an equilateral triangle of side d. It is the circle circumscribed about the triangle. Using

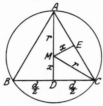

Fig. 51

the notation of Fig. 51 and letting $r + x = h$, we have

$$d^2 = h^2 + \frac{d^2}{4}$$

from the right triangle ABD. This gives us

$$(1) \qquad h^2 = \frac{3d^2}{4}$$

and

$$(2) \qquad h = \frac{d}{2}\sqrt{3}.$$

Now from the triangle DCM we have

$$x^2 + \frac{d^2}{4} = r^2,$$

and hence

$$(h - r)^2 + \frac{d^2}{4} = r^2,$$

$$h^2 - 2hr + r^2 + \frac{d^2}{4} = r^2,$$

$$h^2 + \frac{d^2}{4} = 2hr.$$

Using (1), we obtain

$$d^2 = 2hr,$$

$$r = \frac{d^2}{2h},$$

and then, by (2),

$$r = \frac{d^2}{d\sqrt{3}} = \frac{d}{3}\sqrt{3}.$$

For the equilateral triangle it is evident that the circumscribed circle is the smallest possible enclosing circle. However, we shall not dwell on this point, since it will arise naturally later on.

3. In order to prove Jung's theorem for an arbitrary finite set of points, we will start by trying to choose a circle of least possible radius from all possible enclosing circles. We will use a series of steps that will continually lead us to smaller enclosing circles.

I. An enclosing circle C_1 that has no points of the finite set S on its circumference can always be replaced by a smaller circle C_2. We can draw C_2 with the same center M as C_1, and passing through the point (or points) of S that are farthest away from M.

II. If only *one* point of the finite set S lies on the enclosing circle C_3, then this circle can be replaced by a smaller one (Fig. 52).

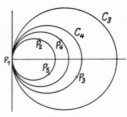

Fig. 52

Letting P_1 be the point of S on the circumference of C_3, we draw all the circles that pass throught another point of S and have the same tangent as C_3 at P_1. These circles all lie inside C_3. Let us call the largest of all these circles C_4. It is different from C_3, since P_1 is the only point of S on the circumference of C_3, while C_4 has another point of S on its circumference. Now C_4 encloses all the new circles and therefore all the points of S. Furthermore, it has *two* of the points on its circumference and it is smaller than C_3.

III. The points of S that lie on the circumference of an enclosing circle divide it into arcs. For the sake of brevity, we shall call an arc "point-free" if no points of S lie on the arc, except that the end points of the arc can be points of S. Our third step for diminishing circles can now be stated: If a point-free arc on an enclosing circle is more than one-half the circumference of the circle, then the enclosing circle can be replaced by a smaller one. [2]

Let P_1 and P_2 be two points of S that are end points of a point-free arc b of the enclosing circle C_5. Furthermore, let b be larger than half the circumference (Fig. 53). We draw the circle C^* with P_1P_2 as diameter. If all the points of S are enclosed by C^*,

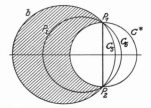

Fig. 53

then it is an enclosing circle smaller than C_5; for C_5, in which the chord P_1P_2 is not a diameter (otherwise b would be one-half a circumference), must be greater than C^*. If not all the points of S are enclosed by C^*, then the remaining points must be in the crescent shaped area between b and C^* (the shaded area of the figure). No points of S other than P_1 and P_2 lie on b, since it is point-free. We now draw all possible circles that pass through P_1, P_2, and a point of S lying in the crescent. The part of the circumference of each of these circles that is inside C_5 is inside the crescent and therefore outside of C^*. The part that is inside C^* is outside of C_5. Let C_6 be the circle whose arc in the crescent extends furthest from the chord P_1P_2. This circle encloses all the points of S, since it encloses all the points of S that lie in the crescent as well as the area common to C_5 and C^*. This area contains all the remaining points of S. Furthermore, C_6 is smaller than C_5. Its circumference lies between the circumferences of C_5 and C^*, so its center is nearer to the chord P_1P_2 than the center of C_5, and hence it is smaller.

If an enclosing circle can no longer be decreased by means of I, II, or III, then it cannot have a point-free arc that is larger than half a circumference. Such a circle must either have two points

[2] Cases I and II can be considered as special cases of III, since in each of them the complete circumference is a point-free arc.

of S as ends of a diameter, or it must have three or more points of S dividing its circumference into arcs that are less than half a circumference. We shall call the first kind of enclosing circle a "diametric circle", the second kind a "three-point circle". The steps I, II, III applied to any enclosing circle will eventually lead to one of these two types. It is possible for an enclosing circle to be of both types at once, for example, a circle through the four vertices of a square.

4. Now we think of having drawn all possible circles that have two points of the finite set as ends of a diameter, as well as all that go through three points of the set. [3] Not all of these circles will be enclosing circles, but every diametric circle and three-point circle will be included. Since altogether there is only a finite number of circles, there can be only a finite number of diametric and three-point circles. Therefore we can compare all these particular circles and pick out the *smallest* one. *This circle c is the smallest possible enclosing circle.* For it is the smallest of all diametric and three-point circles and, by I, II, and III, it must then also be smaller than any other enclosing circle. Furthermore it is *unique.* If there were a second enclosing circle c' of the same size (Fig. 54), then S would lie in c' as well as in c. Then S would lie in the area common to the two circles. Since this whole area can be enclosed by the smaller circle c^*, this would contradict the minimal property of c. We shall call this uniquely determined smallest enclosing circle of the finite set S the *spanning circle of the finite set of points S.*

The spanning circle c can have no point-free arcs of more than half the circumference since, according to III, it would then not be the smallest enclosing circle.

5. Now we shall show that the radius of the spanning circle cannot exceed $\frac{d}{3}\sqrt{3}$. For this purpose we pick out a pair of points of S lying on the circumference of c that are at least as far apart as any other such pair. The distance δ between these points is certainly no greater than the span d of S.

First, it may happen that two points of S are ends of a diameter of c. In this case the diameter $2r$ of c is equal to $\delta \leqq d$. Therefore we have $r \leqq \frac{d}{2}$, and certainly $r < \frac{d}{3}\sqrt{3}$.

---

[3] There are, at most, $\dfrac{n(n-1)}{1\cdot 2} + \dfrac{n(n-1)(n-2)}{1\cdot 2\cdot 3}$ such circles.

Secondly, this may not be the case. Then we pick out the largest point-free arc b on the circumference of c. If there are

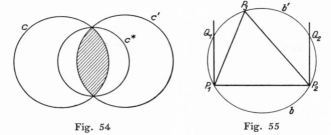

Fig. 54   Fig. 55

several such arcs of equal size, we just pick out any one of them. The end points P_1 and P_2 of b will be points of S. The arc b is *less* than a semicircle, since no point-free arc of c can be greater than half the circumference, and if it were exactly half the circumference P_1 and P_2 would be ends of a diameter and we would be back to the first case. Now we draw the chord P_1P_2 and erect perpendiculars at its ends (Fig. 55). These perpendiculars will cut the circle in two other points, Q_1 and Q_2. The arc b' cut off by Q_1 and Q_2 lies opposite b and is congruent to it. The points Q_1 and Q_2 do not belong to S, for Q_1 and P_2 are ends of a diameter, as are Q_2 and P_1. Hence if Q_1 or Q_2 belonged to S, we would again be back to the first case. However, *the arc b' cannot be point-free*. For if it were it could be extended past Q_1 and Q_2, which do not belong to S_1, to form a larger point-free arc. And this is impossible, since no point-free arc can be larger than b.

Consequently, there is at least one point P_3 of S between Q_1 and Q_2 on b'. The points P_1, P_2, P_3 form an *acute-angle triangle*. The angles at P_1 and P_2 are acute, since they are smaller than the right angles that we constructed at these points. The angle at P_3 intercepts the arc b, which is less than a semicircle. Therefore P_3 is also acute, since an angle inscribed in a semicircle is a right angle, and smaller angles correspond to smaller arcs.

The circumference of c is divided into three arcs by P_1, P_2, P_3. One of these arcs must be at least as large as one-third the circumference, but it is less than a semicircle, since the angles of $P_1P_2P_3$ are acute. Its chord P_iP_j must therefore be at least as great as the chord intercepting one-third the circumference, that is, as great as the side s of an equilateral triangle inscribed in the circle c. Since the length of P_iP_j can at most be the span d, we have $s \leqq d$.

In § 2 we found that the radius r' of the circle circumscribing an equilateral triangle of side s is $r' = \dfrac{s}{3}\sqrt{3}$. Since $s \leq d$ we have, for the radius r of c,

$$r \leq \frac{d}{3}\sqrt{3}$$

which was to be proved.

6. We still want to see whether the bound $\dfrac{d}{3}\sqrt{3}$ cannot be further decreased for general finite sets of points. Let us apply the method of § 4 to the set T consisting of the three vertices of an equilateral triangle. The circles with two points of T as ends of a diameter and the circle through three points of T are drawn in Fig. 56. The only one of these that is an enclosing circle is the one through the three points, the circle circumscribed about the triangle. There are no other circles with which to compare this one, so it is

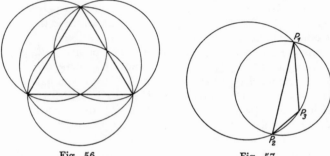

Fig. 56   Fig. 57

the spanning circle. The side of the triangle is the span d and the radius of the circumscribed circle is $\dfrac{d}{3}\sqrt{3}$. Since this particular finite set has a spanning circle of radius $\dfrac{d}{3}\sqrt{3}$, it would be impossible to reduce that bound for general finite sets.

Here we have used § 4 to prove that the circumscribed circle is actually the spanning circle for the set T. That it is the spanning circle seems fairly obvious without a proof, but it is not completely self-evident. The spanning circle of the vertices of an obtuse angle triangle is *not* the circumscribed circle, but the circle having the longest side of the triangle as diameter (Fig. 57).

17. Approximating Irrational Numbers by Means of Rational Numbers

The value $\frac{22}{7}$ is an old and very familiar approximate value for π, the area of a circle of radius 1. Also $\sqrt{2}$ is nearly $\frac{7}{5}$. Precisely what do we mean by such statements as these? The words "approximate" and "nearly" do not have a real place in mathematical speech, yet these statements must have some significance. Why is $\frac{22}{7}$ invariably used to approximate π, in preference, say, to a fraction with denominator 8?

1. If any number w is given, then fractions or (as a mathematician would say) rational numbers that are arbitrarily close to it can be found. For example, if $w = \pi = 3.14159\cdots$, then the fractions

$$3.1 = \frac{31}{10}, \quad 3.14 = \frac{314}{100}, \quad 3.141 = \frac{3141}{1000}, \cdots$$

are getting closer and closer to π. The first fraction clearly differs from π by less than $\frac{1}{10}$, since $\frac{32}{10}$ would already be too large, the second differs by less than $\frac{1}{100}$, etc. In the same way we can approximate any number w to any degree of accuracy if we only know its decimal expansion.

This proposition has a certain esthetic blemish due to its connection with the number 10. Our whole system of decimal notation sets the number 10 apart from all the others, but the choice of 10 is merely a matter of convenience and custom, and it does not reflect any mathematical distinction attached to it. The proposition can be freed from this blemish by being put in the form of the following theorem, where the arbitrary number n takes the place of 10, 10^2, $\cdots$.

Theorem. 1. If w is any number and n is any whole number, then there is a rational number $\frac{m}{n}$, with denominator n, which differs from w by less than $\frac{1}{n}$, $0 \leqq w - \frac{m}{n} < \frac{1}{n}$.

For example, consider $w = \sqrt{2}$, $n = 5$. Then w lies between

1 and 2, and hence it lies in some one of the 5 intervals between the numbers

$$(1) \qquad 1, \ \frac{6}{5}, \ \frac{7}{5}, \ \frac{8}{5}, \ \frac{9}{5}, \ 2.$$

Each of these intervals is of length $\frac{1}{5}$, so all we need do is pick out the last fraction above that is less than $\sqrt{2}$. It will be a fraction with denominator 5 and will differ from 2 by less than $\frac{1}{5}$. The same thing can be done for any arbitrary number w and whole number n. If g is the largest whole number less than w, we consider the rational numbers

$$(2) \qquad g, \ g + \frac{1}{n}, \ g + \frac{2}{n}, \ \cdots, \ g + \frac{n-1}{n}, \ g + 1,$$

and pick out the last one that is less than or equal to w. This number, which we may call $g + \dfrac{l}{n}$, will differ from w by less than $\frac{1}{n}$, so we have

$$(3) \qquad 0 \leq w - \left(g + \frac{l}{n} \right) < \frac{1}{n},$$

which proves the theorem.

Let us find the numbers (1) between which $\sqrt{2}$ falls. The computation is a little simpler if we eliminate the denominator 5. We multiply everything by 5 and look for the position of $5\sqrt{2}$ relative to the numbers

$$(1a) \qquad 5, \ 6, \ 7, \ 8, \ 9, \ 10.$$

Since $5\sqrt{2} = \sqrt{25}\sqrt{2} = \sqrt{50}$, we are really looking for the largest whole number less than $\sqrt{50}$. Now from $49 < 50 < 64$ we have $7 < \sqrt{50} < 8$, and hence 7 is the number we are seeking. Dividing by 5, we see that $\sqrt{2}$ lies between $\frac{7}{5}$ and $\frac{8}{5}$, and we have

$$(4) \qquad 0 \leq \sqrt{2} - \frac{7}{5} < \frac{1}{5}.$$

The general proof of theorem 1 can also be simplified if we

similarly dispense with the denominators. We consider nw instead of w and look for the largest whole number that does not exceed nw. Calling this number m, we have $m \leq nw < m + 1$, and hence $0 \leq nw - m < 1$. Dividing by n, we obtain the required result:

$$0 \leq w - \frac{m}{n} < \frac{1}{n}.$$

2. The fact that there are rational numbers close to π and $\sqrt{2}$ is therefore nothing unusual. What, then, is the significance of the statements that π is approximately $\frac{22}{7}$ and that $\sqrt{2}$ is approximately $\frac{7}{5}$? It lies in the fact that theorem 1 tells us only that $\frac{7}{5}$ will differ from $\sqrt{2}$ by less than $\frac{1}{5}$, while actually it is much closer. It is easy to verify the inequality

$$\frac{7}{5} < \sqrt{2} < \frac{17}{12},$$

and therefore $\sqrt{2}$ differs from $\frac{7}{5}$ by less than $\frac{17}{12}$ does. That is, we have

$$\sqrt{2} - \frac{7}{5} < \frac{17}{12} - \frac{7}{5} = \frac{1}{60},$$

which is much smaller than the $\frac{1}{5}$ guaranteed by theorem 1. Similarly, the inequality

$$3 + \frac{10}{71} < \pi < 3 + \frac{1}{7},$$

which was given by Archimedes, yields

$$\frac{22}{7} - \pi < \left(3 + \frac{10}{70}\right) - \left(3 + \frac{10}{71}\right) = \frac{10}{70} - \frac{10}{71} = \frac{10}{70 \cdot 71} = \frac{1}{497},$$

to be compared with $\frac{1}{7}$ from theorem 1.

Still, this idea of a rational number "much closer" to the given number is not yet a mathematical concept. We are still looking for a definite unequivocal meaning. It is furnished by:

Theorem 2. If w is an irrational number and N is any whole number,

113

then there is a fraction $\frac{m}{n}$, whose denominator does not exceed N, and which differs from w by less than $\frac{1}{nN}$. Furthermore, there is an infinite number of fractions $\frac{m}{n}$ that differ from w by less than $\frac{1}{n^2}$.

Applied to the two previous examples, this asserts that $\sqrt{2} - \frac{7}{5}$ is less than $\frac{1}{5^2} = \frac{1}{25}$ and $\frac{22}{7} - \pi$ is less than $\frac{1}{49}$. Even though these limits still exceed the actual difference, this theorem represents an essential improvement over the first trivial theorem.

To prove the theorem we do not consider the number Nw and the largest whole number less than Nw alone. Instead, we consider the whole series of numbers

$$w,\ 2w,\ 3w,\ \cdots,\ Nw,$$

and the largest whole numbers less than each of these:

$$g_1,\ g_2,\ g_3,\ \cdots,\ g_N.$$

Then we have

$$0<w-g_1<1,\quad 0<2w-g_2<1,\quad 0<3w-g_3<1,\ldots,\quad 0<Nw-g_N<1.$$

The proof will be clearer if we first carry it out for a particular example. We take $w = \sqrt{2}$, $N = 13$ and have

$$
\begin{aligned}
\sqrt{2} &= 1.414\cdots = 1 + 0.414\cdots\\
2\sqrt{2} &= 2.828\cdots = 2 + 0.828\cdots\\
3\sqrt{2} &= 4.242\cdots = 4 + 0.242\cdots\\
4\sqrt{2} &= 5.656\cdots = 5 + 0.656\cdots\\
5\sqrt{2} &= 7.071\cdots = 7 + 0.071\cdots\\
6\sqrt{2} &= 8.485\cdots = 8 + 0.485\cdots\\
7\sqrt{2} &= 9.899\cdots = 9 + 0.899\cdots\\
8\sqrt{2} &= 11.313\cdots = 11 + 0,313\cdots\\
9\sqrt{2} &= 12.727\cdots = 12 + 0.727\cdots\\
10\sqrt{2} &= 14.142\cdots = 14 + 0.142\cdots\\
11\sqrt{2} &= 15.556\cdots = 15 + 0,556\cdots\\
12\sqrt{2} &= 16.970\cdots = 16 + 0.970\cdots\\
13\sqrt{2} &= 18.384\cdots = 18 + 0.384\cdots
\end{aligned}
$$

Of all the amounts by which each of these exceeds the whole number just under it, the fifth is the smallest, $5\sqrt{2} - 7 = 0.071\cdots$, and the 12th is the largest, $12\sqrt{2} = 16 + 0.970\cdots = 17 - 0.030\cdots$ or $17 - 12\sqrt{2} = 0.030\cdots$. We think of taking all 13 of the amounts by which the numbers exceed the whole numbers just under them and putting them in order according to size. Then we have 13 numbers arranged between 0 and 1. They form 14 intervals which may be of various sizes. One of these intervals must have a length of less than $\frac{1}{14}$. For if they all were greater than or equal to $\frac{1}{14}$, then all 14 together would be at least $\frac{14}{14} = 1$ long, and they could have the total length 1 only if each was exactly $\frac{1}{14}$. But if they were all exactly $\frac{1}{14}$, then $\sqrt{2}$, being one of the 13 numbers, would exceed 1 by a whole number of fourteenths, $\sqrt{2} = 1 + \frac{m}{14}$, That is, $\sqrt{2}$ would be rational. Since we have assumed that w is irrational, and have proved the irrationality of $\sqrt{2}$ in Chapter 4, the intervals cannot all be $\frac{1}{14}$, so there must be some interval that is less than $\frac{1}{14}$. We only know that such an interval exists, but not which one it is. However, we can call the lower end of the interval $a\sqrt{2} - g_a = r_a$ and the upper end $b\sqrt{2} - g_b = r_b$. Then we have

$$0 < r_b - r_a = (b\sqrt{2} - g_b) - (a\sqrt{2} - g_a) < \frac{1}{14},$$

and therefore

$$0 < (b - a)\sqrt{2} - (g_b - g_a) < \frac{1}{14}.$$

Since a and b are some two of the numbers 1, 2, 3, $\cdots$ 13, and a and b are unequal, their difference $b - a$ is, apart from sign, again one of these numbers, i.e. $-13 \leq b - a \leq 13$. Therefore $(b-a)\sqrt{2}$ or $(a - b)\sqrt{2}$, whichever is positive, is one of the 13 multiples of $\sqrt{2}$; we shall call it $n\sqrt{2}$. Furthermore $n\sqrt{2}$ differs from the whole number $(g_b - g_a$ or $g_a - g_b)$ just below or just above it by less than $\frac{1}{14}$. Calling this whole number m, we then have

$$-\frac{1}{14} < n\sqrt{2} - m < \frac{1}{14}, \; n \leq 13.$$

Dividing by n we obtain the final result

$$- \frac{1}{14n} < \sqrt{2} - \frac{m}{n} < \frac{1}{14n}.$$

The general proof proceeds in exactly the same way. The irrational number w takes the place of $\sqrt{2}$, $\mathcal{N}$ replaces 13, $\mathcal{N}+1$ replaces 14, and we find that there is some $n \leq \mathcal{N}$ for which we have

$$(5) \qquad - \frac{1}{(\mathcal{N}+1)n} < w - \frac{m}{n} < \frac{1}{(\mathcal{N}+1)n}.$$

This proves the first part of theorem 2.

Since we have $n \leq \mathcal{N}$, formula (5) implies

$$(6) \qquad - \frac{1}{n^2} < w - \frac{m}{n} < \frac{1}{n^2}.$$

Therefore we have found fractions that differ from w by less than $\frac{1}{n^2}$. In (6) the number $\mathcal{N}$ does not appear, but it was used in finding the fraction $\frac{m}{n}$ with denominator $n \leq \mathcal{N}$.

In the previous numerical example n is 5 for $\mathcal{N} = 5, 6, 7, \cdots, 11$, and we have

$$0 < 5\sqrt{2} - 7 < \frac{1}{5}, \quad 0 < \sqrt{2} - \frac{7}{5} < \frac{1}{5^2}.$$

When $\mathcal{N}$ reaches 12, n becomes 12 and we then have

$$0 < 17 - 12\sqrt{2} < \frac{1}{12}, \quad 0 < \frac{17}{12} - \sqrt{2} < \frac{1}{12^2}.$$

For $\mathcal{N} = 13$, n is again 12, etc.

The second part of theorem 2 asserts that for any given irrational w there is an infinite number of fractions $\frac{m}{n}$ that have the property (6). To prove this we need only show that after each fraction $\frac{m}{n}$ satisfying (6), we can always find another one $\frac{m'}{n'}$ that lies still closer to w and still satisfies (6). Since w is irrational, it cannot be equal to any fraction. Therefore

$$w - \frac{m}{n}$$

is not 0 and hence differs from 0 by some definite amount. Because of this there must be a fraction $\dfrac{1}{\mathcal{N}'}$, with $\mathcal{N}'$ sufficiently large, that is closer to 0 than $w - \dfrac{m}{n}$ is. That is, we have

$$(7) \qquad 0 < \frac{1}{\mathcal{N}'} < w - \frac{m}{n} \text{ or } 0 < \frac{1}{\mathcal{N}'} < \frac{m}{n} - w,$$

according to whether $w - \dfrac{m}{n}$ is positive or negative. Using $\mathcal{N}'$ in place of $\mathcal{N}$ in what we have already proved, we can find a fraction $\dfrac{m'}{n'}$, with $n' \leqq \mathcal{N}'$, that satisfies the inequalities

$$(8) \qquad -\frac{1}{(\mathcal{N}' + 1)n'} < w - \frac{m'}{n'} < \frac{1}{(\mathcal{N}' + 1)n'},$$

corresponding to (5). From this we see that the requirement,

$$-\frac{1}{n'^2} < w - \frac{m'}{n'} < \frac{1}{n'^2}$$

is satisfied. Furthermore, since (8) is true, we certainly have

$$-\frac{1}{\mathcal{N}' + 1} < w - \frac{m'}{n'} < \frac{1}{\mathcal{N}' + 1},$$

that is, $w - \dfrac{m'}{n'}$ differs from 0 by *less* than $\dfrac{1}{\mathcal{N}' + 1}$. But according to (7), $w - \dfrac{m}{n}$ differs from 0 by *more* than $\dfrac{1}{\mathcal{N}'}$. Therefore $\dfrac{m'}{n'}$ is closer to w than is $\dfrac{m}{n}$, and consequently $\dfrac{m}{n}$ and $\dfrac{m'}{n'}$ are different.

3. Therefore there is *no last fraction* $\dfrac{m}{n}$ having the property (6). Each one is followed by yet another, so there are infinitely many of them, as theorem 2 asserts. While the trivial theorem 1 admits all whole numbers as denominators, this theorem picks out special denominators with the property (6). We can call these "good denominators." Our earlier numerical example shows that 2, 5, 12 are good denominators for the approximation of $\sqrt{2}$. Further computation reveals the further good denominators 29, 70, 169, $\cdots$.

For π the number 7 is a good denominator. In fact, it is a good deal better than is assured by theorem 2. For we have $\dfrac{22}{7} - \pi < \dfrac{1}{497}$ which is much less than $\dfrac{1}{7^2} = \dfrac{1}{49}$. One might wonder if there are still better denominators than we have found, denominators for which we will have, perhaps, $-\dfrac{1}{n^3} < w - \dfrac{m}{n} < \dfrac{1}{n^3}$ or some similar inequality.

We shall now show that such an improvement of theorem 2 is in general impossible for all irrational numbers w. We shall consider the particular value $w = \sqrt{2}$, and shall show that *every fraction $\dfrac{m}{n}$ differs from $\sqrt{2}$ by more than $\dfrac{1}{3n^2}$.*

We first consider the fractions that are greater than $\sqrt{2}$. For $\dfrac{m}{n} \geqq 2$, since we have $\sqrt{2} < 1.45$, the difference $\dfrac{m}{n} - \sqrt{2}$ is greater than 0.55, while $\dfrac{1}{3n^2} \leqq \dfrac{1}{3}$ is less than 0.55. For $1.55 \leqq \dfrac{m}{n} < 2$ we have $\dfrac{m}{n} - \sqrt{2} \geqq 0.10$, while $\dfrac{1}{3n^2}$ is less than $\dfrac{1}{3 \cdot 4} = \dfrac{1}{12} < 0.10$, since n is now at least 2. Now if $\dfrac{m}{n}$ is between $\sqrt{2}$ and 1.55, we have

$$\left(\frac{m}{n}\right)^2 - 2 = \frac{m^2 - 2n^2}{n^2} = \frac{g}{n^2},$$

where g is some positive whole number. Consequently g is at least 1, and

$$\left(\frac{m}{n}\right)^2 - 2 \geqq \frac{1}{n^2}.$$

Using the formula $a^2 - b^2 = (a + b)(a - b)$, we obtain

$$\left(\frac{m}{n} + \sqrt{2}\right)\left(\frac{m}{n} - \sqrt{2}\right) \geqq \frac{1}{n^2}$$

and then

$$\frac{m}{n} - \sqrt{2} \geqq \frac{1}{n^2} \frac{1}{\dfrac{m}{n} + \sqrt{2}}.$$

Since $\dfrac{m}{n}$ is less than 1.55, we find $\dfrac{m}{n} + \sqrt{2} < 1.55 + 1.45 = 3$, so its reciprocal is greater than $\frac{1}{3}$ and

$$\frac{m}{n} - \sqrt{2} > \frac{1}{3n^2},$$

as was asserted.

Finally, for $0 < \frac{m}{n} < \sqrt{2}$ we can similarly argue:

$$2 - \left(\frac{m}{n}\right)^2 = \frac{g}{n^2} \geqq \frac{1}{n^2},$$

$$\sqrt{2} - \frac{m}{n} \geqq \frac{1}{n^2} \frac{1}{\frac{m}{n} + \sqrt{2}} > \frac{1}{n^2} \frac{1}{2\sqrt{2}} > \frac{1}{3n^2}.$$

18. Producing Rectilinear Motion by Means of Linkages

James Watt's original steam engine was equipped with a remarkable mechanism called Watt's parallelogram because of its shape. The apparatus, which is shown schematically in Fig. 58, consists of five rods hinged together at C, D, E, F. At A and B the rods are held in place by means of pivots that allow the rods to rotate. [1] All the hinges are made in such a way that the rods can never move out of the plane. The piston rod is attached at F. The purpose of this apparatus was to force the end of the piston rod to move along a straight line. This is necessary in order to keep the piston from becoming jammed in the cylinder. However, it can be shown mathematically that the point F does not actually follow exactly a straight line. Rather, it moves along a curve which is so close to being a straight line that the mechanism serves the purpose for which it was intended.

The parallelogram $CDEF$, which gives its name to this linkage, is not really an essential part of it. Its purpose is merely to magnify the usable part of the motion. The essential part is the linkage $ADCB$. For motions that are not too great, this part causes the

---

[1] In the schematic figures, hinges whose positions are fixed will be represented by black circles, movable hinges by hollow circles. Dotted lines are auxiliary lines and do not represent rods.

midpoint *M* of the rod *DC* to approximately follow a
straight line. If the linkage is constructed with

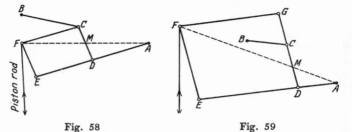

Fig. 58   Fig. 59

$AD = DE = CF = BC$ and $DC = EF$ (the rod *AE* does not bend
at *D*), the points *A*, *M*, *F* will always lie on a line because of the
similarity of the triangles *AEF* and *ADM*. Since, because of the
similarity of the triangles, *AF* is twice *AM*, the point *F* has a motion
that is "similar" to the motion of *M*, but is just twice as great.
The motion of *M* can just as well be magnified *k*-fold. We keep
$AD = BC$ so that the midpoint *M* of *CD* will approximately follow
a straight line, but we increase the size of the parallelogram *DEFG*
(Fig. 59) to obtain

$$AD : AE = DM : EF = 1 : k.$$

In order to investigate the motion that Watt's parallelogram
actually produces, we could now restrict ourselves to a discussion
of the motion of *M*. It is easy to show that the motion is not
rectilinear by considering several extreme positions of the linkage.
However, since this particular motion is not of importance to our
other topics, we shall not study it further.

2. The production of straight-line motion by means of linkages
has been of considerable importance in the history of machine
design, and the problem has been taken up by a large number of
workers. But the particular use to which Watt put his linkage is
not important any longer. On a modern steam engine the end of the
piston rod is held in a straight line by an entirely different sort of
mechanism, a "crosshead" that slides between parallel rails. It
may be that the early designers were led to use linkages rather than
the crosshead principle because they had an incorrect idea of the
size of the frictional forces involved.

The linkage problem has not only occupied practical designers,
but it has also attracted the attention of pure mathematicians.

The mathematicians have naturally considered the problem in its strictest form, to find a linkage in which one pivot will follow a theoretically exact straight line. The first to take up this problem was the great Russian mathematician P. L. Tschebyschev (1821–1894), who studied Watt's parallelogram and its possible improvements without finding a linkage that would produce accurately a straight line. Many other fruitless attempts were made during the first half of the last century, and finally mathematicians began to doubt whether an exact solution were possible.

Then, in 1864, Peaucellier devised a linkage that produces straight-line motion. This apparatus is called Peaucellier's cell. Then, as so often happens in the history of discovery, a great many solutions of the problem were found. Linkages were found that would produce various curves of which the straight line is only a particular case.

3. We shall now consider Peaucellier's cell. This linkage produces not only straight-line motion but also an "inversion" of the plane. An inversion is a particular kind of "mapping" of the plane onto itself. By a mapping we mean an picturing of the plane in which each point has an image. Some simple mappings are: (1) reflection in a line, which we used repeatedly in Chapters 5 and 6; (2) parallel displacement, in which the image of each point is found by displacing the point itself by a fixed amount in a fixed direction; (3) the rotation of the plane about a fixed point through a fixed angle; (4) the stretching of the plane about a point (the center), in which each point is moved out along the ray from the center so that its distance from the center is increased by a fixed ratio $1 : \lambda$. The image of a point need not always be different from the original point. In parallel displacement, every point is mapped on an image that is different from it, but in rotation or stretching the center has itself as its image. In the case of reflection in a line, each point of the line is mapped onto itself.

These examples should serve to explain what we mean by a mapping. Now we must see just what type of mapping an *inversion* is. A circle C with center O, the center of the inversion, is given in Fig. 60. To find the image P' of any point P, we draw a ray from O through P. This ray will cut the circle at some point Q. We then determine P' on the ray to satisfy $OP : OQ = OQ : OP'$. If we call the radius of the circle a and write $OP = r$, $OP' = r'$, then the proportion can be replaced by the equation

$$rr' = a^2.$$

From the definition of an inversion, it is clear that the image of every point inside the circle of inversion C is outside the circle and

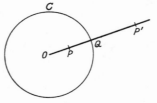

Fig. 60

vice versa. Each point of the circle of inversion is its own image. Because of this, an inversion is often called "reflection in a circle". Just as with ordinary reflection, inversion has the property that the image of the image is the original point itself.

4. So far as single points are concerned, all the properties of an inversion are immediately evident from the definition. New questions arise if we consider the images of curves, but we cannot undertake to make a complete investigation of this question. The most important result is that every line or circle is mapped by an inversion into either a line or a circle. We shall require only a part of this statement, namely, that *the image of a circle that passes through the center of inversion is a line.* We shall prove this assertion.

Let k be the circle passing through the center O, and let its diameter be $OA = d$ (Fig. 61). The circle of inversion K has radius a. The image A' of A lies on the extension of the diameter OA, and the distance $OA' = d'$ must satisfy the equation $dd' = a^2$. Now we

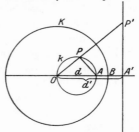

Fig. 61

shall prove that the image of k is the line perpendicular to OA' through A'. To do this we must show that every line through O meeting the circle k at a point P meets the line at P', with $OP \cdot OP' = a^2$.

If we draw the line AP, we have two right triangles OAP and $OP'A'$ which are similar because they have a common angle at O. Therefore we have the proportion

$$OP : OA = OA' : OP',$$

or, with $OP = r$, $OP' = r'$,

$$rr' = dd'.$$

But using $dd' = a^2$, we have

$$rr' = a^2,$$

and this is just the relation between a point and its image required for an inversion.

Beside this fact, we shall also need a result that is proved in elementary geometry. If two secants are drawn through a fixed point outside a circle, the product of one and its external segment equals the product of the other and its external segment. In the notation of Fig. 62, this theorem can be written as $s_1 s_1' = s_2 s_2'$. Its proof follows directly from the similarity of the triangles AP_1P_2' and AP_2P_1', which have the common angle A and in which the angles at P_1' and P_2' are equal because they intercept the same arc P_1P_2.

5. With these preliminaries out of the way, we may now return to Peaucellier's cell (Fig. 63). It consists of four rods of length c, hinged to form a rhombus $PQP'R$, and two rods of length $b > c$,

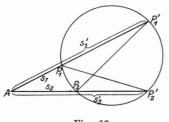

Fig. 62

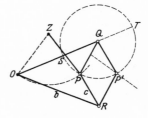

Fig. 63

hinged to two opposite vertices Q and R of the rhombus. The other ends of these two rods pivot about the fixed point O.

Because of the symmetry of the system of rods, the points O, P, P' are always on a line, the axis of symmetry of the figure. We draw a circle about Q with radius c. This circle passes through P and P' and crosses the rod OQ and its extension in two points S and T. Then OPP' and OST are two secants of the circle, to which we can apply the elementary theorem mentioned above. It gives us the result

$$OP \cdot OP' = OS \cdot OT = (b - c)(b + c) = b^2 - c^2.$$

Therefore the product $OP \cdot OP'$ is a constant. If we set $a^2 = b^2 - c^2$, then a is one leg of a right triangle with hypotenuse b and other leg c. The equation $OP \cdot OP' = a^2$ then shows that P' is the image of P under an inversion in the circle of radius a about O.

Peaucellier's cell therefore sets up an inversion in the part of the plane that can be reached by P and P'. To obtain rectilinear motion, we need only cause P to follow a circle passing through O. Then, from what we know of inversions, the image P' will move along a straight line. There is no difficulty in forcing P to follow a circle. We merely hinge a rod at P and pivot its end at a *fixed* point Z. In order to make the circle pass through O we must make the length of the rod ZP equal to the fixed distance OZ.

With the added rod, Peaucellier's cell consists of 7 rods. The point P is restrained to an arc of a circle which, perhaps only when extended, passes through O, and P' then moves along a straight line. This line will be perpendicular to OZ.

6. Other linkages which produce inversions, and which can then be used to produce rectilinear motion (just as with Peaucellier's cell), have been found. One of these, devised by Hart, requires only 4 rods instead of the 6 of Peaucellier's cell. The 4 rods are hinged together as they would be to form a parallelogram $ABCD$, but the parallelogram is "turned inside out". That is, the points A and C are pulled apart until the parallelogram collapses into a straight line, and then D and B are brought to the same side of AC (Fig. 64). The two "diagonals" AC and DB are always parallel because the triangles ACB and ACD are congruent and therefore have equal altitudes. Now the linkage is held in any *one* position

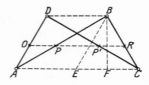

Fig. 64

and the points O, P, P', R are marked on the 4 rods on a line parallel to these diagonals. We shall now show that *these points always lie on a line parallel to the diagonals,* regardless of the position of the linkage.

If we first look at the triangle DAB in Fig. 64, we notice that OP is parallel to DB in the original position of the linkage. For this reason, we have

$$(1) \qquad AO : OD = AP : PB.$$

We have proved (1) only for the original position of the linkage, but since it is a proportion involving only the lengths of segments, it must remain true for *every* position of the linkage. From (1) we now have $OP \parallel DB$ for every position of the linkage. Now, looking at the triangle ADC, we see that OP' is parallel to AC in the original position of the linkage, so we have

$$(2) \qquad DO : OA = DP' : P'C.$$

This proportion between lengths of segments remains true for all positions of the linkage, so we have $AC \parallel OP'$. Since we have proved $OP \parallel DB$, $DB \parallel AC$, $AC \parallel OP'$, we have $OP \parallel OP'$, and OP and OP' are the same line since they are parallel and have the point O in common. This line is parallel to DB. Similar reasoning applied to the triangle DCB shows that $P'R \parallel DB$ and we find that R also lies on the line containing O, P, P'.

Furthermore, because the three lines are parallel, we have

$$OP : DB = AO : AD,$$
$$OP' : AC = DO : DA.$$

These proportions can be replaced by the equations

$$(3) \qquad OP \cdot AD = AO \cdot DB,$$

$$(4) \qquad OP' \cdot DA = DO \cdot AC.$$

If we multiply corresponding sides of these equations together and divide by AD^2, we find

$$(5) \qquad OP \cdot OP' = \frac{AO \cdot DO}{AD^2} \cdot AC \cdot DB.$$

On the right side the numbers AO, DO, AD are fixed lengths that are determined by the apparatus, while AC and DB are the lengths of the diagonals and apparently depend on the position of the linkage. However, *the product $AC \cdot DB$ is a constant.* For, drawing $BE \parallel DA$ and $BF \perp AC$, we have

$$(6) \qquad AC \cdot BD = (AF + FC)(AF - FC) = AF^2 - FC^2.$$

125

From the right triangles AFB and FCB and the Pythagorean theorem we find

$$AF^2 + FB^2 = AB^2,$$
$$FC^2 + FB^2 = CB^2,$$

and then, by subtraction,

$$AF^2 - FC^2 = AB^2 - CB^2.$$

This, with (6), gives

(7) $$AC \cdot BD = AB^2 - CB^2,$$

and the right side is a constant that does not depend on the position of the linkage. Now from (7) and (5) we have

(8) $$OP \cdot OP' = \frac{AO \cdot DO}{AD^2} (AB^2 - CB^2),$$

where the right side is a constant that can be determined by making certain measurements on the linkage. Since 0, P, P' lie on a line, the point P' will be the image of P under an inversion if we keep the position of O fixed. The circle of inversion has its center at O, and the square of its radius is given by the right side of (8).

The proof that Hart's linkage produces an inversion is now complete. In order to produce rectilinear motion, we need only add a fifth rod that will restrain P to a circle through O.

7. There are also linkages that produce rectilinear motion without depending on an inversion. The double rhomboid of Kempe (Fig. 65) is a particularly ingenious arrangement. By the term "rhomboid" we mean a four-sided figure in which two pairs of adjacent sides are equal. In a rhomboid the two angles formed by two unequal sides are always equal. Let $ABCD$ and $BCEF$ be two hinged rhomboids in which the ratio of the smaller

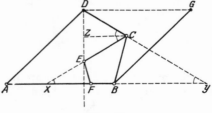

Fig. 65

side to the larger is the same in both, and in which the larger side of the smaller rhomboid is equal to the smaller side of the larger rhomboid. Then in Fig. 65 we have $AD = AB$, $CD = CB = CE$, $FB = FE$ and $CB : AD = FB : CB$. Since the vertex F of the smaller rhomboid lies on one side of the larger, the two rhomboids have the angle at B in common. Therefore the angles at B, at E, and at D are all equal. Now two convex quadrilaterals are similar if their sides are in the same ratio and they have one pair of corresponding angles equal. No matter how the linkage is moved, the two rhomboids $ABCD$ and $BCEF$ will always remain similar.

If two opposite sides of a rhomboid are extended until they intersect, they form an angle. These angles: EXF and CYB are evidently equal for similar rhomboids. If we draw $CZ \parallel AB$ through C, we have $\angle DCZ = \angle CYB$, since they are corresponding angles of parallel lines. We also have $\angle ZCE = \angle EXF$, since they are alternate interior angles. Therefore all four of these angles, which are marked in the figure, are equal. Consequently CZ is the bisector of the vertex angle of the triangle CDE, which is isosceles by hypothesis, and it is perpendicular to the base DE. Since CZ was drawn parallel to AB, the line DE will then *always* be perpendicular to AB.

Now if we keep D fixed and move AB in such a way that it always remains parallel to its original position, then the perpendicular from D to AB will remain fixed. But the point E is on this perpendicular, so it can have only *rectilinear motion* along this perpendicular. In order to keep AB always parallel to itself, we add a rod BG of length $BG = AB = AD$. We keep the end G of the rod fixed at the point determined by the condition $DG = AB$. Then $ABGD$ is a rhombus, so its opposite sides AB and DG remain parallel for all positions of the linkage. That is, AB always remains parallel to the *fixed* line DG and hence is always parallel to itself. Thus rectilinear motion is finally attained.

8. The double rhomboid of Kempe can be arranged to give rectilinear motion in another especially elegant way. We supply ourselves with two copies $ABCDEF$ and $A'B'C'D'E'F'$ of the double rhomboid linkage, one of which is the mirror image of the other. We connect these two so that that they have the point C in common, and so that DCE' and $D'CE$ become rigid rods that do not bend at C (Fig. 66). The bisector CZ of angle DCE is, as we have seen, parallel to AB, and similarly the bisector CZ' of angle $D'CE'$ is parallel to $A'B'$. Since the angles DCE and $D'CE'$ are vertical angles by construction, the two bisectors CZ and CZ' lie in a line. The lines AB and $A'B'$ are parallel to this line, so they are parallel

to each other for every position of the linkage. More than this, the line $A'B'$ always lies on the extension of the line AB. Since

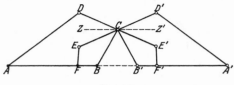

Fig. 66

we know that the two lines are parallel, all we need do is prove that the point C is equidistant from each of these lines AB and $A'B'$. We shall do this by showing that the two rhomboids $ABCD$ and $A'B'CD'$ are always congruent for all positions of the linkage. In the first place, since the angles DCE and $D'CE'$ are vertical angles, they are equal. By § 7 these angles are twice the angle between two opposite sides in each of the rhomboids. Corresponding sides of the two rhomboids are equal. If we can show that a rhomboid is completely determined by the lengths of its sides and the angle between a pair of opposite sides, then our two rhomboids will be congruent, since they are determined by equal sides and an equal angle. To see that the rhomboid is completely determined we construct the rhombus $BCDH$ inside the rhomboid $ABCD$ (Fig. 67). The points A, H, C lie on a line because of symmetry. [2] Since HB

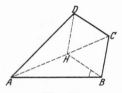

Fig. 67

is parallel to DC, the angle ABH is equal to the angle between two opposite sides. It is therefore determined as half the given angle DCE of Fig. 66. Since AB is given and BH is equal to the given BC, the triangle ABH is completely determined. Then D is determined as the mirror image of B in AH, and C is determined as the fourth vertex of the rhombus $BCDH$, three of whose vertices, B, H, D, have already been determined.

Therefore, as was asserted, the two halves of Fig. 66 are always

[2] This figure is exactly that of Peaucellier's cell.

congruent to one another. Consequently the height of C above AB equals its height above $A'B'$, and then AB and $A'B'$ lie on the same line.

Now if we keep A and B fixed, the rod $A'B'$ can move but must remain on the line AB. Thus we have obtained rectilinear motion from this new linkage.

This linkage does considerably more than the previous ones. In the previous mechanisms there is a single point that moves along a line. In this case a whole rod $A'B'$ moves along the line in which it lies. Since any figure can be rigidly attached to $A'B'$, this apparatus accomplishes a parallel displacement of any figure or even of the whole plane by which all points move along equal and parallel line segments.

19. Perfect Numbers

Book IX of Euclid's *Elements* is the third and last book that is occupied with arithmetic. This book includes the proof of the infinitude of prime numbers, which we have reproduced in Chapter 1, and it concludes with a discussion of so-called perfect numbers. Perfect numbers are also mentioned by Plato, especially in an enigmatic passage in his *Republic* where, in an obscure discussion of eugenics, he introduces the "nuptial number".

The subject of perfect numbers and the theorems that were later proved concerning them are now no more than an interesting curiosity in the body of modern mathematics. But we shall discuss this minor topic because in the method Euclid used there burns the tiny spark of an idea which, as we shall see in the next chapter, was rekindled by Euler and has flared up into a great flame, the modern theory of the distribution of prime numbers.

Euclid defined a perfect number as one that is equal to the sum of all its divisors. For example, the number 6 is a perfect number, since its divisors are easily found by trial to be 1, 2, 3, and $1+2+3=6$. In this definition of a perfect number, the number itself is obviously not counted as one of its divisors. By continued trials, the next perfect number after 6 is found to be $28 = 1 + 2 + 4 + 7 + 14$. The next perfect number is too large to be easily found by trial, but this is unimportant. What we really want is a general systematic method of finding perfect numbers.

1. It is clear that a prime number cannot be a perfect number. 1 and p are all the divisors of the prime p, and p itself is not to be counted. Therefore the sum of all the divisors is merely 1, which is certainly not equal to p.

2. Having taken care of this very simple case, we can consider one only a little more complicated. The number 9 is not a prime, but it is the square of a prime. By trying the numbers from 1 through 8, we see that 1 and 3 are all the divisors that are to be counted. Since $1 + 3 = 4$ is less than 9, the number 9 is not perfect.

Correspondingly, the numbers 1 and p are obviously divisors of p^2, the square of a prime. Now we cannot verify by trial, as in the case of 9, that these are all the divisors that are to be counted. We can, however, *prove* that these two divisors exhaust the possibilities. This was done by Euclid after he had established the necessary lemmas. Our proof is basically the same as Euclid's except for the way in which it is formulated. It is based on the theorem of the unique factorization into prime factors, which we proved in Chapter 11. If p^2 had any other divisors, it could be factored in some way different from $p^2 = p \cdot p$. In fact the other factorization would involve some prime different from p, and this is impossible because of the unique factorization theorem.

3. More generally, we can show in the same way that no power of a prime can be a perfect number. Just as for p^2, we see that all the divisors of p^a that are to be counted are

(1) $$1, \ p, \ p^2, \cdots, \ p^{a-1}.$$

Since they form a geometric series, their sum is given by a familiar formula, one that Euclid proved expressly for the present purpose. In fact,

(1a) $$1 + p + p^2 + \cdots + p^{a-1} = \frac{p^a - 1}{p - 1}.$$

Because the denominator $p - 1$ is 1 for the smallest prime $p = 2$, and is larger for all larger primes, the fraction in (1a) is never any larger than its numerator $p^a - 1$. Therefore the sum of the divisors of p^a is never any larger than $p^a - 1$. It is less than p^a, and hence p^a is not a perfect number.

4. Now we can see how to discuss a number like $72 = 2^3 \cdot 3^2$, which contains two primes. We shall immediately consider the number $p^3 q^2$. The divisors of $p^3 q^2$, including the number $p^3 q^2$ itself, are

$$1, \ p, \ p^2, \ p^3,$$
$$(2) \qquad q, \ qp, \ qp^2, \ qp^3,$$
$$q^2 \ q^2p, \ q^2p^2, \ q^2p^3,$$

that is, all numbers of the form $p^\alpha q^\beta$ where α goes up to 3, β to 2. It is obvious that all of these numbers divide p^3q^2. It again follows directly from the theorem on unique factorization that these include *all* the divisors. The table (2) gives us a complete view of all the divisors of our number. In order to find the sum of the divisors, we note that the second row is merely the first row multiplied by q, and that the third row is the first multiplied by q^2. The sum of the first row is the sum of the geometric series $1 + p + p^2 + p^3$, so the complete sum is this sum taken 1 time plus q times plus q^2 times. Therefore the complete sum is

$$D = (1 + q + q^2)(1 + p + p^2 + p^3).$$

Analogously, the sum of the divisors of $N = p^a q^b$ is

$$D = (1 + q + q^2 + \cdots + q^b)(1 + p + p^2 + \cdots + p^a)$$
$$(2a) \qquad = \frac{q^{b+1} - 1}{q - 1} \, \frac{p^{a+1} - 1}{p - 1},$$

but in these formulas we must remember that we have included the number N itself among the divisors.

5. This last result can be generalized to numbers N containing more than two primes. For $N = p^a q^b r^c$, besides a table like (2), we will also have the whole table multiplied by r, and by r^2, and so on until it is finally multiplied by r^c. In all, the table is multiplied by $(1 + r + r^2 + \cdots + r^c)$. If N has still another prime factor, then the sum of the divisors has another corresponding factor. In general, for $N = p^a q^b r^c \cdots$, the sum of the divisors of N is

$$D = (1 + p + p^2 + \cdots + p^a) \ (1 + q + q^2 + \cdots + q^b) \ (1 + r + r^2 + \cdots + r^c) \cdots$$
$$(3) \qquad = \frac{p^{a+1} - 1}{p - 1} \frac{q^{b+1} - 1}{q - 1} \frac{r^{c+1} - 1}{r - 1} \cdots,$$

where, again, we are including the number N itself.

6. Basically all of this is included in Euclid, but it is given expressly only for the case $N = p \cdot 2^b$. In this case, where N is a product of two factors, one a power of 2, the other a prime p to the first power, formula (2a) becomes

$$D = \frac{2^{b+1} - 1}{2 - 1} \cdot \frac{p^2 - 1}{p - 1}.$$

If N is a perfect number, we have $D = 2N$, since N is included among the divisors, and

$$D = (2^{b+1} - 1)(p + 1) = 2N.$$

Putting in the value of N, we find

$$(2^{b+1} - 1)(p + 1) = 2 \cdot p \cdot 2^b = 2^{b+1}p,$$
$$2^{b+1}(p + 1) - (p + 1) = 2^{b+1}p,$$
$$2^{b+1} = p + 1,$$
(4) $$p = 2^{b+1} - 1.$$

Therefore if p is not only a prime but is also equal to $2^{b+1} - 1$, then N is a perfect number. The number $2^{b+1} - 1$ is not always a prime for every value of b, but we have Euclid's theorem:

The number $N = (2^{n+1}-1) \cdot 2^n$ is a perfect number for all numbers n for which $2^{n+1} - 1$ is a prime.

7. Let us try the values $n = 1, 2, 3, \cdots$ in order to find some further perfect numbers:

$$n = 1, \quad N = (2^2 - 1) \cdot 2 = \quad 3 \cdot 2 = 6,$$
$$n = 2, \quad N = (2^3 - 1) \cdot 2^2 = \quad 7 \cdot 4 = 28$$
$$n = 3, \quad N = (2^4 - 1) \cdot 2^3 = \quad 15 \cdot 8 \text{ not perfect, } 15 \text{ not prime,}$$
$$n = 4, \quad N = (2^5 - 1) \cdot 2^4 = \quad 31 \cdot 16 = 496,$$
$$n = 5, \quad N = (2^6 - 1) \cdot 2^5 = \quad 63 \cdot 32 \text{ not perfect, } 63 \text{ not prime,}$$
$$n = 6, \quad N = (2^7 - 1) \cdot 2^6 = 127 \cdot 64 = 8128,$$

. . . .

Euclid's theorem immediately gives rise to a new problem: *For what values of n is $2^{n+1} - 1$ a prime?* *One* step towards its solution can be made at once. If $n + 1$ is a composite number, say $n + 1 = uv$, we have $2^{n+1} - 1 = 2^{uv} - 1 = (2^u)^v - 1$. Now, by the formula for the sum of a geometric series, we find

$$x^v - 1 = [x^{v-1} + x^{v-2} + \cdots + x + 1][x - 1],$$

and letting $x = 2^u$ we obtain

$$(2^u)^v - 1 = [(2^u)^{v-1} + \cdots + (2^u) + 1][(2^u) - 1],$$

a product of two factors. Therefore $2^{n+1} - 1$ cannot be a prime unless $n + 1$ is a prime itself. In the continuation of the above table, the next prime after 7 is 11, so the next value that must be tried is $n = 10$. Since $2^{11} - 1 = 2047 = 23 \cdot 89$ is not a prime, the number $(2^{11} - 1) \cdot 2^{10}$ is not a perfect number, but the factorization of 2047 has required a very considerable number of trials. Several further numbers have been found to be perfect, but no complete rule for finding them all has ever been discovered. Another question that has remained unanswered is whether there is an infinite number of these perfect numbers or whether there is finally a last one.

8. Euclid's result only goes as far as this theorem, but there is the most varied evidence that more was known at that time. For one thing, Jamblichos states without further explanation that there are no even perfect numbers other than those given by Euclid. It is not known whether this was proved by the ancients or, if so, how they did it. However, a proof based on the formula (3) has been given by Euler, and this formula is essentially contained in Euclid's work. Even though it is not important to the following topics, we shall reproduce this interesting proof.

Let N be any even number. Then 2 is one of the prime factors of N and it appears to some power, say 2^n. The remaining prime factors of N are odd, and together they represent some odd number u. Then we have $N = 2^n u$. If we form the sum of the divisors D of N according to (3), the first factor is

$$\frac{2^{n+1} - 1}{2 - 1} = 2^{n+1} - 1,$$

and the second factor is exactly what would arise if (3) were used to find the sum U of the divisors of u. We now have

(5) $$D = (2^{n+1} - 1)U,$$

which must equal $2N$ if N is perfect, since we have included N among the divisors. Therefore if N is perfect we find

$$(2^{n+1} - 1)U = 2N = 2 \cdot 2^n u = 2^{n+1}u,$$
$$2^{n+1}U - U = 2^{n+1}u,$$
$$2^{n+1}(U - u) = U = (U - u) + u,$$
(6) $$(2^{n+1} - 1)(U - u) = u.$$

Now U is the sum of the divisors of u including u itself, so $U - u$ is the sum of the divisors of u, *not* including u itself. Also, (6) implies that $(U - u)$ divides u if N is a perfect number, that is, $(U - u)$ is a divisor of u. If $U - u$ is a divisor itself and is at the same time the sum of all the divisors, then it must be the *only* divisor of u. Since 1 is certainly a divisor of u, it must be this single divisor $U - u$, and u must be a prime. Then (6) simplifies to $u = 2^{n+1} - 1$, and we have $N = 2^n(2^{n+1} - 1)$. Every even perfect number is of Euclid's form.

9. At this point we encounter a second question: *Are there any odd perfect numbers?* This problem also remains unsolved. No one has found an odd perfect number, and it appears very unlikely that one exists, but this has not been proved.

10. A passage in the fifth book of Plato's *Laws* also leads one to think that Euclid did not include all that was known at his time. There Plato recommends that, in a newly-founded city, the number of plots of land and of landowners be chosen so that it will have as many divisors as possible, perhaps 5040 with $60-1$ divisors. He points out that the legislators must understand enough arithmetic to be able to arrange for cities of any size. This involves the problem of the *number* of divisors of a number. In order to solve it, we can use part of our discussion concerning the *sum* of the divisors. For example, *all* the divisors of p^3q^2 are included in the table (2) and we can count them by rows, $3 \cdot 4 = 12$. This number includes the number p^3q^2 itself, so we must subtract 1 to obtain the number $12-1$ of proper divisors. More generally for $N = p^aq^b$ the number of proper divisors of N is

$$P = (a + 1)(b + 1) - 1,$$

and for the completely general case $N = p^aq^br^c \cdots$, it is

$$P = (a + 1)(b + 1)(c + 1) \cdots - 1.$$

In particular, for $N = 5040 = 2^4 \cdot 3^2 \cdot 5 \cdot 7$ we have

$$P = (4 + 1)(2 + 1)(1 + 1)(1 + 1) - 1 = 5 \cdot 3 \cdot 2 \cdot 2 - 1 = 60 - 1.$$

This presumably explains Plato's use of the unusual and clumsy notation $60-1$ instead of 59; and this notation, along with his instructions that the legislators must know the answer to this problem, leads one to believe that he was in possession of the solution.

11. The close connection between the number of divisors and the sum of the divisors of a number becomes more obvious if we think of them as two special cases of a general topic. This is the sum S of the sth powers of the divisors of a number N. For example, for $N = 6$ we have $S = 1^s + 2^s + 3^s$. For $s = 1$ the sum S is the sum D of proper divisors of N, while for $s = 0$ each individual term is 1 and S is the number P of proper divisors of N. For $s = 2$ it would be the sum of the squares of the proper divisors of N, and so on. For each value of s, a formula analogous to (3) can be found by the same argument. The value $s = -1$ may be used as well as any other. The $-$1st powers are the reciprocals of the divisors; so, for $N = 6$, S is

$$\frac{1}{1} + \frac{1}{2} + \frac{1}{3}.$$

The formula analogous to (3) for the sum of the reciprocals of the divisors of N is clearly

7) $\quad R = \left(1 + \frac{1}{p} + \cdots + \frac{1}{p^a}\right)\left(1 + \frac{1}{q} + \cdots + \frac{1}{q^b}\right)\left(1 + \frac{1}{r} + \cdots + \frac{1}{r^c}\right) \cdots$

All of these sums are part of a single theory which arises from the fact that a table like (2) gives us a complete list of all the divisors of a number. The whole basis of the theory was known to the ancients, and the evidence in Plato seems to indicate that they recognized its beauty and importance.

20. Euler's Proof of the Infinitude of the Prime Numbers

Euclid's proof of the infinitude of the primes, which we discussed in Chapter 1, immediately precedes his consideration of perfect numbers. Euler, who took up and extended the study of perfect numbers, produced another proof of the infinitude of primes. This proof uses the same ideas that are basic in the theory of perfect numbers.

We must make two simple observations before proceeding with the proof.

1. Let AB be a line segment 2 feet long.

$A \qquad\qquad\qquad\qquad M \qquad\quad M_1 \quad M_2 \quad B$

If we traverse it from A to its midpoint M, from there to the midpoint M_1 of the remainder MB, from there to the midpoint M_2 of the remainder M_1B, etc., we always remain short of B, but we continually come closer and closer to B. We will have covered distances of 1 ft., $\frac{1}{2}$ ft., $\frac{1}{4}$ ft., and so on. All these distances together will total less than 2 ft.:

$$1 + \frac{1}{2} + \frac{1}{2^2} + \cdots + \frac{1}{2^n} < 2.$$

If the segments continually decrease in some other ratio $x < 1$

instead of $\frac{1}{2}$, the same thing is true. To prove this, we again use the formula for the sum of a geometric series,

$$1+x+x^2+ \cdots +x^n= \frac{1-x^{n+1}}{1-x} = \frac{1}{1-x} - \frac{x^{n+1}}{1-x}.$$

Since x is less than 1, the second fraction on the right is a positive quantity that is subtracted from the first fraction, so we have

$$1 + x + x^2 + \cdots + x^n < \frac{1}{1-x}.$$

If p is any prime number, we have $\frac{1}{p} < 1$, and we can replace x by $\frac{1}{p}$ in our inequality. We can then assert that *if p is any prime and n is a whole number*, then

$$(1) \qquad 1 + \frac{1}{p} + \frac{1}{p^2} + \cdots + \frac{1}{p^n} < \frac{1}{1-\frac{1}{p}} = \frac{p}{p-1}.$$

2. If we write

$$A_m = 1 + \frac{1}{2} + \frac{1}{3} + \cdots + \frac{1}{2^m}$$

then

$$1+ \frac{1}{2} + \frac{1}{3} + \frac{1}{4} > 1 + \frac{1}{2} + \left(\frac{1}{4} + \frac{1}{4}\right)= 1 + \frac{2}{2},$$

$$1+ \frac{1}{2} + \frac{1}{3} + \frac{1}{4} + \frac{1}{5} + \frac{1}{6} + \frac{1}{7} + \frac{1}{8} > A_2+\left(\frac{1}{8} + \frac{1}{8} + \frac{1}{8} + \frac{1}{8}\right) > 1+\frac{3}{2},$$

$$1+ \frac{1}{2} + \frac{1}{3} + \cdots + \frac{1}{16} > A_3 + \left(\frac{1}{16} + \cdots + \frac{1}{16}\right) > 1+\frac{4}{2},$$

and, in general,

$$(2) \qquad A_m = 1 + \frac{1}{2} + \frac{1}{3} + \cdots + \frac{1}{2^m} > 1 + \frac{m}{2}.$$

The larger m is, the larger A_m becomes, and A_m can be made as large as we please by taking m large enough. For example, we need only choose $m = 1998$ in order to have $\frac{m}{2} = 999$ and hence $A_m > 1000$. In the same way, we can choose m large enough to make A_m greater than one million if desired. [1]

3. We let p be any prime and write (1) for each of the primes 2, 3, 5, $\cdots$, up to p:

$$1 + \frac{1}{2} + \frac{1}{2^2} + \cdots + \frac{1}{2^n} < \frac{2}{2-1},$$

$$1 + \frac{1}{3} + \frac{1}{3^2} + \cdots + \frac{1}{3^n} < \frac{3}{3-1},$$

$$\cdots\cdots\cdots\cdots\cdots\cdots\cdots\cdots\cdots$$

$$1 + \frac{1}{p} + \frac{1}{p^2} + \cdots + \frac{1}{p^n} < \frac{p}{p-1}.$$

Now we form the product R_p of all these sums. This product will be less than the product M_p of all the right sides,

$$R_p < M_p = \frac{2}{1} \cdot \frac{3}{2} \cdot \frac{5}{4} \cdot \frac{7}{6} \cdots \frac{p}{p-1}.$$

We are already familiar, from the previous chapter, with products of the sort R_p. When multiplied out, R_p is the sum of all the terms

$$\frac{1}{2^\alpha} \cdot \frac{1}{3^\beta} \cdots \frac{1}{p^\mu},$$

where each of α, β, $\cdots$, μ takes on all values up to n. In other words, R_p is the sum of the reciprocals of all the divisors of

$$N = 2^n \cdot 3^n \cdot 5^n \cdots p^n.$$

Therefore R_p is a sum of reciprocals of whole numbers, but not of all possible whole numbers. The only numbers that occur are those made up of the prime factors 2, 3, 5, $\cdots$, p, and each of these primes appears to no higher than the nth power. For example,

$$\left(1+\frac{1}{2}+\frac{1}{2^2}\right)\left(1+\frac{1}{3}+\frac{1}{3^2}\right) = 1+\frac{1}{2}+\frac{1}{3}+\frac{1}{4}+\frac{1}{6}+\frac{1}{9}+\frac{1}{12}+\frac{1}{18}+\frac{1}{36}$$

[1] It follows from this that $1 + \frac{1}{2} + \frac{1}{3} + \cdots + \frac{1}{n}$ increases without bound as n increases (here n is not restricted to be a power of 2, but as it increases it will eventually exceed any determined power of 2). In mathematics this is expressed by saying that the "infinite series $1 + \frac{1}{2} + \frac{1}{3} + \cdots$ diverges" even though its individual terms continually get smaller and smaller. However, this important fact and the idea of convergence and of infinite processes is not involved in the present topic.

is the sum of the reciprocals of all divisors of $2^2 \cdot 3^2 = 36$.

4. We now come to the actual proof. Let m be any positive whole number. Then A_m is the sum of the reciprocals of all the numbers,

$$(3) \qquad\qquad 1, \; 2, \; 3, \; 4, \cdots, \; 2^m.$$

We consider all the primes that divide all the numbers (3), and call the largest of these primes q. Then all the numbers (3) are products involving only the prime factors 2, 3, 5, $\cdots$, q. Furthermore, none of these primes can appear to a power higher than the mth power. For if the smallest prime 2 appears in a number to a power higher than the mth, then that number is larger than 2^m and hence does not appear in the series (3). What is true for the prime 2 is certainly true for all the larger primes. Therefore *the numbers* (3) *are all included among the divisors of* $2^m \cdot 3^m \cdot 5^m \cdots q^m$.

Now if we form the product R_p of § 3 not for an arbitrary p but for our particular prime q, and with m instead of n, it will include *all* the terms of A_m. The expression R_q will contain other terms than those of A_m, but the important thing is that all the terms of A_m are to be found among those of R_q. In connection with the example at the end of § 3, we see that *all* the terms of $A_2 = 1 + \dfrac{1}{2} + \dfrac{1}{3} + \dfrac{1}{4}$ are included, along with several other terms.

Because of this relation between A_m and R_q we have $A_m < R_q$. Furthermore, we have $R_q < M_q$ by § 3, and $1 + \dfrac{m}{2} < A_m$ from (2). Combining these, we find

$$(4) \qquad\qquad 1 + \frac{m}{2} < \frac{2}{1} \cdot \frac{3}{2} \cdot \frac{5}{4} \cdots \frac{q}{q-1}.$$

Now m was an arbitrary positive whole number and q was the largest of all the prime factors of all the numbers in (3). The left side of (4) can be made arbitrarily large by merely choosing m large enough. If there were only a finite number of primes, q could increase up to the last prime but no further. Then the right side of (4) would increase to a certain value and would then remain constant. This would contradict the fact that the left side can be made arbitrarily large; therefore it is proved that the primes are infinite in number.

This proof is far more complicated than that of Chapter 1. But its importance lies in the fact that the same methods can be used for a great many similar but more difficult problems, a few of which were mentioned in Chapter 1. These methods form the basis for the theory of the distribution of primes, one of the most extensive and difficult fields of modern mathematics. At least in essence, they have been handed down to us from the ancients.

21. Fundamental Principles of Maximum Problems

We have repeatedly discussed maximum problems. In Chapters 3, 5, and 6 we exhibited some mathematical miniatures which mathematicians of great ability have found time to produce along with their more important and lengthier work. In this chapter we shall discuss some principles that are basic to all these problems.

These principles can be developed by considering an extremely simple maximum problem. *Let a triangle be given* (it is best to think of it as cut out of paper). The problem is to *find the two points P and Q that are as far apart as possible on the surface or its boundary* (Fig. 68). The answer is easy to guess: P and Q are the ends of the longest side. But how can we prove this?

There is a simple method, a recipe, that we have not used in the earlier chapters. It will lead us to the answer here as well as in the other problems. We argue as follows: If one of the points, say P, lies on the *inside* of the triangle, then PQ certainly does not have its

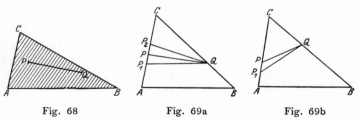

| Fig. 68 | Fig. 69a | Fig. 69b |

maximum length. For on the extension of the line PQ there is a point P_1 that is further from Q than P is, and that is still inside the triangle. If both P and Q lie on the boundary of the triangle, but one of them, say P, is not a vertex, then we can find a nearby point P_1 on the boundary that is further from Q than the distance PQ. That this is so is clear when PQ is perpendicular to the side on which P lies (Fig. 69a), as well as when PQ is not perpendicular to this side (Fig. 69b). Therefore PQ can be a maximum only if both

P and Q are vertices; otherwise it certainly is not. Then PQ is a side of the triangle and must naturally be the longest side.

The same argument will give us the corresponding result for a polygon: *in order that two points on the surface of a polygon be farthest apart, they must be two of the vertices that are farthest apart.* In this result, the polygon does not need to be convex. In the quadrilateral of Fig. 70 the two vertices at the bottom are the points that are farthest apart.

The maximum problems in the previous chapters can be handled in the same way. This principle leads to the result much more

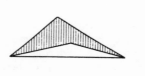

Fig. 70

Fig. 71

quickly than the methods given in those chapters. For example, to find the largest triangle inscribed in a given circle (Fig. 71), we suppose that ACB is not equilateral. If AC and BC are unequal we inscribe in the circle the isosceles triangle ABD, with base AB. The new triangle has a greater altitude than the original, so it has a greater area. Therefore a triangle that is not equilateral cannot be a maximum; the maximum triangle must be equilateral. The same methods also apply in the case of the pedal triangle.

Why did we use very much longer and more complicated proofs in the previous chapters? Why wouldn't we have been satisfied with this much shorter method? The reason is that this procedure has a serious logical defect. This defect remained unnoticed for two centuries until it was brought out clearly by Weierstrass in the second half of the nineteenth century.

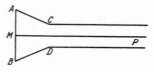

Fig. 72

The defect will become apparent if we apply the principle to another example (Fig. 72). Because the plane has infinite extent, it is clear that there are no two points that are farthest apart in this figure; the farther the point P moves to the right, the greater becomes the distance MP. However, the previous argument can be made just as before. Suppose P and Q lie anywhere inside the figure or on the boundary, even including the possibility that they may be at any of the four vertices A, B, C, D. Unless PQ is exactly the side AB, a nearby point P_1 can be found that increases the distance PQ. Just as in the earlier cases, for each pair of points P, Q we can find a nearby pair that are further apart in every case except when the pair is A, B. No pair other than A, B can give a maximum. If we now follow the previous argument strictly, we must conclude that AB is the maximum.

In this last case we have obtained an obviously incorrect result by following exactly the same principles. How can we be sure that the answer we obtained for the triangle is correct?

The defect in the method is this: in the case of the triangle we proved conclusively that no pair of points, other than the ends of the longest side, can possibly be a maximum. But this does not tell us that these points are farther apart than any other pair in the triangle. If we know that there is a solution of the maximum problem, we can logically conclude that it is this pair of points, since there is no other possibility. But we must know that there is a solution. In the case of Fig. 72, if there were a solution it would have to be AB, but here there actually is no maximum.

It is now clear why we were unable to avoid the apparently cumbersome proofs in the earlier chapters. They were not only desirable for aesthetic reasons, but they were necessary to avoid a serious logical error as well. In order to complete all the topics of this chapter we must still supply the lacking proof for Fig. 68.

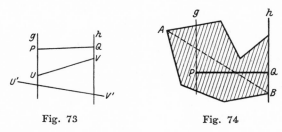

Fig. 73   Fig. 74

We first make a very elementary observation: if we have two parallel lines g and h (Fig. 73), then the perpendicular PQ is at least as short as any other segment UV between a point on one line and a point on the other. Also, PQ is shorter than any segment $U'V'$ that joins a point U' on the left of g with a point V' on the right of h.

Instead of limiting ourselves to a triangle, we can just as easily carry out the proof for a general polygon.

Let P and Q be any two points inside the polygon or on the boundary. We draw the two perpendiculars g, h to the line PQ at its ends (Fig. 74). These two lines cut a parallel strip out of the whole figure. Now, since at least the point P of the polygon lies on g, some vertex A of the polygon must lie to the left of g, or possibly on g. Also, some other vertex B must lie to the right of, or possibly on, h. The distance AB is not less than PQ, and therefore PQ is not greater than the largest of all the distances between any two vertices of the polygon. This last distance is therefore a maximum.

In the following chapter we shall take up a maximum problem which is considerably more difficult than any of the previous ones. It will be concerned with finding the figure, bounded by lines or curves and with a given perimeter, which has the largest area. We shall prove that the only possible maximum is a circle, but shall not attempt to prove that the circle actually has a greater area than any other figure. This problem is complicated by the fact that curved figures come into consideration. Because of this we will have to use a whole series of new ideas and facts in order to obtain even this limited result.

22. The Figure of Greatest Area with a Given Perimeter

Why do soap bubbles have the shape of a sphere? It is because the walls are made of a substance that is subject to cohesive forces tending to increase the thickness and decrease the area of the walls. The pressure of the air does not come into play, but the enclosed air maintains a fixed volume while the area becomes as small as possible. The soap bubble solves the problem of finding the solid figure with given volume that has the least area.

The problem that we shall solve is more modest than the one which is solved by every soap bubble and every raindrop. Instead of considering solid figures, we shall restrict ourselves to two dimensions and ask for the figure of least perimeter with a given area.

1. Here we have asked for the figure of least *perimeter* with a given area, while the title of this chapter asks for the figure of greatest *area* with a given perimeter. But this difference is only apparent. To make this clear, suppose that it is not the circle K with given perimeter k (Fig. 75), but the figure F with given perimeter

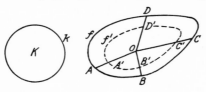

Fig. 75

f that has a maximum area. Then we have $f = k$ and $F > K$. Now we can contract the figure F proportionally by choosing any point O inside of it, drawing lines $OA, OB, OC, \cdots$ from O to the boundary, and then diminishing each of these lines by a fixed ratio. We can choose this ratio so that the contracted figure F', with perimeter f', has the same area as K. This diminishes the perimeter, so we have $f' < k$, and therefore F' is a figure with the same area as the circle but a smaller perimeter. If this is proved to be impossible, then the problem in the title is also solved and the solution is the circle. Since the converse can also be proved in the same way, the two formulations are equivalent.

2. We have implicitly used the following theorem concerning the perimeter and area of a figure:

I. *If a figure is contracted proportionally around a point O in the ratio* $1 : r$, *then the perimeter is diminished in the same ratio* $1 : r$ *and the area is diminished in the ratio* $1 : r^2$.

We shall use several other theorems concerning perimeters and areas of figures. We shall not take up the question of how these are proved. This would require an analysis of the exact meanings of all the concepts involved, and then we should have to build up a complete theory. It is not our purpose to do this, but we shall list here all the theorems which we shall use without giving their proofs.

II. *If the surface of one figure is part of the surface of another, then it has a smaller area than the other* (Fig. 76).

The analogous statement for the perimeter of a figure is *not* true. This can be seen from Fig. 77, where we merely need to make the inner curve twist enough so that it is longer than the outer curve. However, the theorem becomes correct if we limit it to *convex* figures.

A figure is called convex if every two points of the figure can be joined by a straight line that lies entirely in the figure. A circle and a triangle

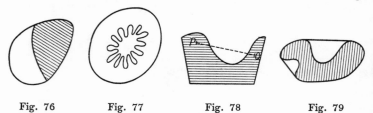

Fig. 76   Fig. 77   Fig. 78   Fig. 79

are convex figures, while Fig. 78 is not, since P and Q are in it but the line joining them is not entirely in the figure.

III. *If one convex curve encloses another convex curve, then the enclosing curve has the larger perimeter* (Fig. 76).

IV. *Every non-convex figure can be rounded off to a convex figure with a larger area and smaller perimeter* (Fig. 79).

These four theorems which we shall accept without proof are all intuitively self-evident, with the possible exception of III. However, we have learned that intuitive evidence is not of much importance.

3. We are now ready to begin the proof, following the argument used by J. Steiner. We first prove that *the curve of greatest area with a given perimeter must be convex.* This follows directly from IV and I. If we round the curve off according to IV, we obtain a convex figure with smaller perimeter and larger area. If we now expand proportionally by the appropriate ratio to obtain the original perimeter., the area is again increased, as can be seen from I. We have constructed a convex curve with the same perimeter but with a greater area. Therefore a curve that is not convex can never have maximum area for a given perimeter.

4. We need only consider convex figures from now on. Next we will prove that *corresponding to any convex figure there is a convex figure of the same perimeter, with an area at least as large, and which has an axis of symmetry.*

Let P be any point on the curve (Fig. 80), and let Q be the point on the curve whose distance from P, measured along the curve, is exactly half the perimeter. The chord PQ cuts the figure into two parts. If these parts have different areas we choose the larger part and reflect it in the line PQ; if they have equal areas, we choose either part and reflect it in PQ. This part and its reflection form a figure with the same perimeter as the original and with area at least

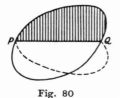

Fig. 80   Fig. 81

as large. Furthermore, this figure has an axis of symmetry. If the figure is not convex, it can be made so by using IV and I, as was done in § 3. This will only serve to increase the area, and it is easy to see that it will not destroy the symmetry.

This completes the proof of the assertion, but there is still another point to be mentioned. If the original figure is a circle, then PQ is a diameter and the figure obtained by reflection is the same circle. If the original figure is not a circle, then at least one of the two parts into which PQ divides it is not a semicircle (Fig. 81). If one part is larger than the other, then the new figure obtained by reflection has a larger area. If the two parts have equal areas, let us always agree to choose the part that is *not* a semicircle. Then the new figure will not be a circle. If the new figure is not convex, we can round it off by means of IV and I, once more increasing its area. Therefore we can now say either that there is a figure of greater area with the same perimeter, or that the new figure is convex and has the same area and perimeter as the original. Furthermore, the new figure will be a circle only if the original one is circular.

5. If the original figure is not a circle, we can either "better" it or replace it with a convex figure with the same area and perimeter, a figure which possesses an axis of symmetry but is not a circle. If the figure can be "bettered", it is certainly not a maximum.

We will now show that if a convex figure with an axis of symmetry is not a circle, we can construct another convex figure with the same perimeter and with an area that is definitely larger. This will complete the proof, since we will then have proved that every figure that is not a circle can be "bettered".

If AB is the axis of symmetry (Fig. 82), there must be some point C on the bounding curve that makes the angle ACB *different* from a right angle. For if every point formed a right angle, then, according to a well-known theorem (see Fig. 18), the curve would be a circle with diameter AB, and we are supposing that the curve is not a circle.

We now construct a right triangle $A_1B_1C_1$ (Fig. 83) with legs $A_1C_1 = AC$ and $B_1C_1 = BC$. We reflect this triangle in A_1B_1, thus forming a quadrilateral $A_1C_1B_1C_1'$. On the sides of this quadrilateral we place the corresponding segments of the original figure that are cut off by $ACBC'$ (the segments that are shaded in Figs. 82 and 83).

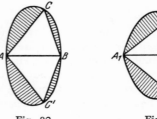

Fig. 82   Fig. 83

The perimeter for the new figure remains the same as for the old, since it consists of the same four arcs. The area of the new figure is made up of the four segments and the quadrilateral $A_1C_1B_1C_1'$. The segments are the same in both figures, so we need only compare the quadrilaterals or even the triangles ABC and $A_1B_1C_1$, since the figures are symmetrical. The triangle $A_1B_1C_1$ has the area $\frac{1}{2}A_1C_1 \cdot B_1C_1$. The altitude of ABC from B is *less* than BC, since the angle at C is not a right angle. Therefore the area of ABC is *less* than $\frac{1}{2}AC \cdot BC$. Now we have $A_1C_1 = AC$ and $B_1C_1 = BC$, so the area of ABC is *less* than that of $A_1B_1C_1$, and therefore the new figure has a larger area than the original.

The proof is now complete: if we are given any figure that is not a circle, we can find another figure with the same perimeter and a larger area. The only figure that can possibly be a maximum is the circle.

We have not proved that the circle has a greater area than any other curve with the same perimeter. The relation between what we have proved and the complete problem has already been discussed in the last chapter. The complete solution of this problem can be given, but it would necessitate the systematic building-up of an extensive theory, and that is beyond the intentions of this book.

23. Periodic Decimal Fractions

1. The expansion of a common fraction into a decimal is a familiar process.   When it is carried out, decimals of quite different sorts may arise, as is shown by the following examples:

I. $\frac{1}{5} = 0.2$, $\frac{3}{40} = 0.075$,

II. $\frac{4}{9} = 0.4444\cdots$, $\frac{1}{7} = 0.142857142857\cdots$,

III. $\frac{1}{6} = 0.1666\cdots$, $\frac{7}{30} = 0.2333\cdots$.

The simplest are those of type I.   Remembering the meaning of a decimal, we can write them as common fractions with their denominator powers of 10:

$$\frac{1}{5} = \frac{2}{10}, \quad \frac{3}{40} = \frac{75}{1000}.$$

These equations do not imply anything unusual.   They merely assert that the given fractions can be "extended" so that their denominators become powers of 10.   That can be done with any fraction whose denominator divides some power of 10.   The characteristic of such a denominator is that it has no prime factors other than 2 and 5, since $2^\alpha \cdot 5^\beta$ will divide 10^γ if γ is the larger of α and β.   Therefore a fraction with the denominator $2^\alpha \cdot 5^\beta$ can always be extended to a fraction with denominator 10^γ, and consequently to a γ-place decimal.

2. If the denominator of a fraction has a factor k which is not divisible by 2 or 5, then it cannot be extended to a fraction with a power of 10 as the denominator.   For the assumption

$$\frac{1}{2^\alpha \cdot 5^\beta \cdot k} = \frac{a}{10^\delta} = \frac{a}{2^\delta \cdot 5^\delta}$$

would imply

$$2^{\delta-\alpha} 5^{\delta-\beta} = a \cdot k.$$

The number k, which is supposed to be greater than 1, would therefore be divisible by 2 or 5 since, because of unique factorization, 2 and 5 are the only primes dividing $a \cdot k$ and hence k.

The examples II and III are of this type.   In this case we say that the fractions $\frac{4}{9}$, $\frac{1}{7}$, $\frac{1}{6}$, and $\frac{7}{30}$ can be expanded into "infinite decimal fractions."   These are not, properly, decimal *fractions* in the sense of having a denominator that is a power of 10.   There is no last digit in the decimal expansion, so there is no suitable power of 10.

When we speak of an "infinite decimal fraction" we are using the word "decimal fraction" in a new and extended sense. For our purposes there is no point in going into the exact meaning of this new sense. All that concerns us is the formal side of the question, in this case the process of expanding the fraction into a decimal does not break off. It unendingly produces more and more digits of the infinite decimal. More than this, the infinite decimals will be *periodic*. That is, from a certain point on, the sequence of digits will consist of the mere repetition, over and over again, of the same group of digits. For example, the expansion of $\frac{1}{7}$ consists, from the very start, of the repetition of 142857. Since there are only 10 digits, some digit must appear repeatedly in any infinite decimal fraction, but this does not mean that all such decimals are periodic. An example of a non-periodic infinite decimal fraction is the decimal

$$0.101001000100001\ldots,$$

where only the digits 0 and 1 are used, and where the nth 1 is followed by exactly n digits 0.

3. Examples of fractions in which the denominator has a prime factor other than 2 and 5 are given in II and III. We shall now show that such fractions always lead to periodic decimal expansions. First considering the example $\frac{1}{7}$, we obtain the decimal expansion by the usual method of long division:

$$
\begin{array}{r}
0.142857\ldots \\
7\,)\,\overline{1.000000} \\
7 \\
\overline{30} \\
28 \\
\overline{20} \\
14 \\
\overline{60} \\
56 \\
\overline{40} \\
35 \\
\overline{50} \\
49 \\
\overline{1}
\end{array}
$$

As soon as the remainder 1 has appeared, the whole process of division starts over again right from the beginning. The sequence of remainders 1, 3, 2, 6, 4, 5 and the sequence of quotients 1, 4, 2, 8, 5, 7 will repeat themselves over and over. The decimal fraction will have the period 142857. Since we will want to carry out the divisions for other examples, it will be convenient to represent the process in a more compact way. We shall write

$$\frac{1}{7} = 0.\overline{142857} \cdots$$
$$\phantom{\frac{1}{7} = 0.}{\scriptstyle 1\ 3\ 2\ 6\ 4\ 5\ 1}$$

where we have written the remainder under the newly-found quotient at each step. The period of the infinite decimal fraction will always be distinguished by a bar over it.

Another example is

$$\frac{3}{41} = 0.\overline{0\ 7\ 3\ 1\ 7} \cdots.$$
$$\phantom{\frac{3}{41} = 0.}{\scriptstyle 3\ \ 30\ 13\ 7\ 29\ 3}$$

Here the division process starts to repeat when the remainder 3 appears, since the whole process began with 3. Some of the digits in the quotient can repeat before the end of the period; it is not this but the repetition of a digit in the *remainder* that causes the process to begin over again and the period to end.

In carrying out the division for any fraction $\frac{a}{b}$ we must eventually find a remainder that has already appeared before. (The numerator of the fraction is considered as the first remainder.) The only remainders that are available are 1, 2, 3, $\cdots$, $b - 1$. The remainder 0 is not included, since its appearance would mean that the division comes out even and the decimal is finite. But we have seen that this is impossible if the denominator has a prime factor other than 2 and 5. Now, since there are only $b - 1$ remainders available, the division process must start over anew at the bth step or earlier. The decimal expansion of $\frac{a}{b}$ is periodic and its period has at most $b - 1$ places. For $\frac{1}{7}$ this maximum length is reached, for $\frac{3}{41}$ it is not reached. A further example with period of maximum length is

$$\frac{1}{17} = 0.\overline{0\ 5\ 8\ 8\ 2\ 3\ 5\ 2\ 9\ 4\ 1\ 1\ 7\ 6\ 4\ 7} \cdots$$
$$\phantom{\frac{1}{17} = 0.}{\scriptstyle 1\ \ 10\ 15\ 14\ 4\ 6\ 9\ 5\ 16\ 7\ 2\ 3\ 13\ 11\ 8\ 12\ 1}$$

with period of length $b - 1 = 17 - 1 = 16$.

4. From now on we shall consider fractions whose denominators are relatively prime to 10, that is, those which do not have

the prime factors 2 or 5. We will then be able to say considerably more about the length of the period.

The fraction $\dfrac{a}{b}$ will naturally be a *reduced* fraction; a and b are relatively prime. Then the remainders that appear in the division will all be prime to b. For suppose r is a remainder that is prime to b. The next step of the division consists of dividing $10r$ by b to obtain a quotient q and a remainder r_1. That is, b goes into $10r$ just q times and it leaves over the remainder r_1,

$$10r = qb + r_1$$

or

(1) $$r_1 = 10r - qb.$$

No prime divisor of b can divide either 10 or r. Therefore none divides $10r$ and none divides $10r - qb$. Consequently every remainder r that is prime to b is followed by another remainder r_1 that is also prime to b. Since the numerator a is counted as the first remainder and it is prime to b, so will all the remainders be prime to b. Therefore *the period of* $\dfrac{a}{b}$ *can be no longer than the number of remainders that are prime to b.*

The number of remainders prime to b is a number which is of interest for its own sake. In the theory of numbers it is usually designated by the symbol $\varphi(b)$, and we have, for example, $\varphi(2)=1$, $\varphi(3) = 2$, $\varphi(4) = 2$, $\varphi(5) = 4$, $\varphi(6) = 2$. $\varphi(7) = 6$, $\varphi(8) = 4$. Since for a prime p, all the $p - 1$ smaller numbers are prime to it, we have

(2) $$\varphi(p) = p - 1.$$

Using the symbol $\varphi(b)$, we can now say that the period of $\dfrac{a}{b}$ has at most $\varphi(b)$ places if b is prime to 10.

5. We have seen that the decimal expansion for a fraction $\dfrac{a}{b}$ must be periodic if b has a prime factor different from 2 and 5, even when b is not prime to 10. We saw that some one of the finite number of possible remainders must eventually be repeated. Therefore we knew that there must be a period, but we could not be sure where the period would begin. In the example II, the period begins immediately after the decimal point, while in III it begins only after having passed a digit. Now the denominators of II were chosen to have no factor in common with 10, and we will show that the period of the decimal expansion of $\dfrac{a}{b}$ will always

begin immediately after the decimal point if b is prime to 10. To do this we will have to show that the first remainder that is repeated is the first in the whole series, the numerator itself. If two remainders are equal, $r_m = r_n$, then the two preceding remainders r_{m-1} and r_{n-1} are also equal, if there actually are any preceding remainders. For r_m and r_n have arisen from the division of $10r_{m-1}$ and $10r_{n-1}$,

$$10r_{m-1} = q_{m-1} \cdot b + r_m,$$
$$10r_{n-1} = q_{n-1} \cdot b + r_n.$$

Using $r_m = r_n$, we find by subtraction

$$10(r_{m-1} - r_{n-1}) = (q_{m-1} - q_{n-1})b,$$

and therefore b divides $10(r_{m-1} - r_{n-1})$. Since b is prime to 10 it must divide $r_{m-1} - r_{n-1}$, so this difference must be one of the numbers,

$$0, \pm b, \pm 2b, \pm 3b, \cdots.$$

Now, however, r_{m-1} and r_{n-1} are remainders, so each is less than b, and this difference is then numerically less than b. The only possibility is that

$$r_{m-1} - r_{n-1} = 0,$$
$$r_{m-1} = r_{n-1}.$$

Therefore the period begins as early as possible, immediately after the decimal point.

6. If we designate the length of the period by λ, then in developing $\frac{a}{b}$ in a decimal fraction we come to the first remainder a again after λ steps of the division process. At each step of the division we have moved one more place to the right of the decimal. The remainder a that appears after λ steps really represents $\frac{a}{10^\lambda}$. Therefore if we divide $10^\lambda a$ by b the remainder is a, and consequently $10^\lambda a - a$ is divisible by b. Since we have

$$10^\lambda a - a = a(10^\lambda - 1),$$

and since a is prime to b, this means that b divides $10^\lambda - 1$. Since the period ends at the *first* repetition of the remainder a, λ is the *smallest* number for which $10^\lambda - 1$ is divisible by b, that is, the smallest number for which $a \cdot 10^\lambda$ has the remainder a on division by b.

We have proved

Theorem I.   *The length λ of the period of $\dfrac{a}{b}$ is the smallest number λ for which $10^{\lambda} - 1$ is divisible by b.*

Two facts that are implicitly included in this theorem deserve to be emphasized.   *First,* to every number b that is relatively prime to 10, there corresponds a number λ such that $10^{\lambda} - 1$ is divisible by b.   The existence of such a λ is not self-evident, but we have obtained it as the length of the period and have proved the existence of a period in § 3.   *Second,* the number λ is determined by b and is *independent* of a.   All reduced fractions $\dfrac{a}{b}$ with the same denominator b have periods of the same length.   We shall emphasize the dependence of λ on b by writing $\lambda = \lambda(b)$.

7.   We shall now consider the decimal expansions of $\dfrac{a}{b}$ for different values of a but the same b.   We have already found

$$\frac{1}{7} = 0.\overline{142857}\cdots,$$
$$\phantom{\frac{1}{7} = 0.}{\scriptstyle 1\ 3\ 2\ 6\ 4\ 5\ 1}$$

and similarly obtain

$$\frac{2}{7} = 0.\overline{285714}\cdots.$$
$$\phantom{\frac{2}{7} = 0.}{\scriptstyle 2\ 6\ 4\ 5\ 1\ 3\ 2}$$

The period 285714 belonging to $\frac{2}{7}$ can be obtained from the period 142857 of $\frac{1}{7}$ by a cyclic reordering.   This is clear if we notice that the remainder 2, with which the development of $\frac{2}{7}$ begins, appears also among the remainders for $\frac{1}{7}$.   From this point on the two developments must agree step for step.   The remainders 3, 4, 5, 6 also arise in the development of $\frac{1}{7}$.   In fact all the 6 remainders for 7 must arise since the development of $\frac{1}{7}$ has a period of length 6.   If we think of the digits of the period as a cycle in which the last digit of the cycle is supposed to be followed by the first digit of the period, then we can arrange the remainders and quotients in a table:

remainders	1	3	2	6	4	5
quotients	1	4	2	8	5	7

From the table we can read off the decimal expansion of $\frac{6}{7}$ or any other similar fraction.   For example, the period of $\frac{6}{7}$ begins with 8 and is therefore 857142, so we have $\frac{6}{7} = 0.\overline{857142}$.

The periods that we have been speaking of are periods of quotients.   As we have seen, however, the remainders also exhibit periods of the same sort and the same length.

8. The periods of $\frac{a}{b}$ cannot exceed $\varphi(b)$ in length, but it is not necessary that they actually be this long. For $\frac{1}{7}$ and $\frac{1}{17}$ we found periods of this maximum length, but for $\frac{3}{41}$ we found $\lambda = 5$, which is less than $\varphi(41) = 40$. The actual length of the period is difficult to determine in advance and it must be found by computation in each case. But we can say more about its length $\lambda(b)$ than the single result $\lambda(b) \leqq \varphi(b)$.

We shall use the new example $\frac{1}{21}$ as a starting point for our discussion. By testing the 20 numbers less than 21 we easily find $\varphi(21) = 12$. However, by division we obtain

$$\frac{1}{21} = 0 . \overline{0\,4\,7\,6\,1\,9} \cdots,$$
$$\phantom{\frac{1}{21} = 0 . }\scriptstyle 1 \quad 10\ 16\ 13\ 4\ 19\ 1$$

with period of length $\lambda(21) = 6 < 12 = \varphi(21)$. In the division process only 6 of the 12 possible remainders arise and we arrange them in a table:

(A)

remainders	1	10	16	13	4	19
quotients	0	4	7	6	1	9

From this table we can read

$$\frac{10}{21} = 0.\overline{476190} \cdots, \qquad \frac{4}{21} = 0.\overline{190476} \cdots,$$

and others, but we cannot find the expansion of $\frac{2}{21}$. This fraction requires a new division,

$$\frac{2}{21} = 0 . \overline{0\,9\,5\,2\,3\,8} \cdots,$$
$$\phantom{\frac{2}{21} = 0 . }\scriptstyle 2 \quad 20\ 11\ 5\ 8\ 17\ 2$$

from which we obtain a second table:

(B)

remainders	2	20	11	5	8	17
quotients	0	9	5	2	3	8

Not only does the one new remainder 2 appear in this table, but all of the other remainders are new as well. It might have been seen before that none of the old remainders would appear. Each remainder completely determines the whole further course of the development, so every remainder of table (A) carries with it the entire 6-digit period of remainders of (A). Then, if (B) contained any remainder which appears in (A), it would contain all of (A). Since (B) has only 6 remainders, it could not contain any others; but this does not agree with the fact that (B) contains the remainder 2. The two tables (A) and (B) together now contain all the $\varphi(21) = 12$ remainders that can be numerators of reduced proper fractions with denominator 21.

9. The ideas involved in the construction of tables (A) and (B) for the denominator 21 can be used to obtain some general results concerning the length of the period.

If the development of $\dfrac{1}{b}$ has the maximum length $\lambda(b) = \varphi(b)$, there is just one table, as in the case of $\frac{1}{7}$ in § 7.

However, if we find $\lambda(b) < \varphi(b)$, as for $b = 21$, then only $\lambda(b)$ remainders occur in the development of $\dfrac{1}{b}$. Using these we form a table (A) which clearly does not contain all the $\varphi(b)$ possible remainders. We choose a remainder r which is prime to b and does not appear in (A), and develop $\dfrac{r}{b}$ in a decimal fraction whose period will also have the length $\lambda(b)$, according to § 6. We use the new remainders and quotients to form a second table (B). The remainders in (B) will all be different from those in (A), since (B) contains r and any remainder of (A) would carry with it all $\lambda(b)$ remainders of (A).

The tables (A) and (B) together contain $2\lambda(b)$ different remainders, all prime to b. Either this number represents all the possible remainders, in which case we have $2\lambda(b) = \varphi(b)$, or there are still other remainders. If s is a remainder which is not in (A) or (B), we develop $\dfrac{s}{b}$ and obtain another table (C), containing $\lambda(b)$ new remainders which are neither in (A) nor (B). In all, we would now have $3\lambda(b)$ different remainders, prime to b. If $3\lambda(b)$ is equal to $\varphi(b)$, all the possible remainders have been exhausted. Otherwise we repeat the process, forming new tables until all $\varphi(b)$ possible remainders are used. The important thing is that each time a new remainder appears, there are $(\lambda - 1)$ other new ones to go along with it.

This tabulating will finally result in a number, say k, of tables which contain all $\varphi(b)$ remainders prime to b. Each table contains $\lambda(b)$ remainders and no remainder occurs twice. Therefore

(3) $$\varphi(b) = k \cdot \lambda(b),$$

which proves the theorem:

The length $\lambda(b)$ of the period is a divisor [1] of $\varphi(b)$.

We have already found $\varphi(p) = p - 1$ if p is prime. Therefore we have the special result that the length $\lambda(p)$ of the period for the fraction $\dfrac{a}{p}$ is a divisor of $p - 1$ if p is a prime. A number of examples of this may be found above, where we found $\lambda(3) = 1$, $\lambda(7) = 6$, $\lambda(17) = 16$, $\lambda(41) = 5$.

As one more example of the method, we will consider the distribution of the remainders among the various periods for the denominator 39. We start with

(A)
$$\frac{1}{39} = 0.\overline{025641}\cdots,$$
$$\scriptstyle 1\ \ 10\,22\,25\,16\,41$$

from which the table of remainders and quotients can easily be constructed. The question then arises whether the table contains all the remainders that are prime to 39. Evidently the remainder 2 is missing, so we use it for the construction of the next table:

(B)
$$\frac{2}{39} = 0.\overline{051282}\cdots.$$
$$\scriptstyle 2\ \ 20\,5\,11\,32\,8\,2$$

This yields 6 new remainders. The smallest remainder which is prime to 39 and which is not in (A) or (B) is 7. This allows us to find a new table:

(C)
$$\frac{7}{39} = 0.\overline{179487}\cdots.$$
$$\scriptstyle 7\ \ 31\,37\,19\,34\,28\,7$$

Now 14 is the smallest remainder prime to 39 that is not included among the 18 remainders of (A), (B), and (C), and we use it to continue our process:

(D)
$$\frac{14}{39} = 0.\overline{358974}\cdots.$$
$$\scriptstyle 14\ \ 23\,35\,38\,29\,17\,14$$

In all we now have 24 different remainders, all prime to 39. These finally exhaust all the possible remainders. In fact, among the numbers from 1 to 39 the multiples of 3 and 13 will have a common factor with 39. The multiples of 3 account for one-third of all these numbers, 13 in all, and the multiples of 13 account for 2 more, the numbers 13 and 26 (39 has already been counted as a multiple of 3). In all, 15 of the numbers from 1 to 39 have a common factor with 39, and the remaining $39 - 15 = 24$ are prime to 39. Therefore we have $\varphi(39) = 24$, which is in agreement with $\lambda(39) = 6$

---

[1] The improper divisor $\varphi(b)$ itself is not to be excluded.

Furthermore, since $\varphi(39) = 4 \cdot \lambda(39)$, there are 4 tables, (A), (B), (C), and (D).

10. Our results concerning the length of the period can be used to obtain an important general theorem. We need the following simple lemma:

If x and k are positive whole numbers, then $x^k - 1$ is divisible by $x - 1$.

This lemma follows directly from the formula

$$(1 + x + x^2 + \cdots + x^{k-1})(x - 1) = x^k - 1,$$

which can be obtained from the formula for the sum of a geometric series, or by direct multiplication. If we use the value $x = 10^{\lambda(b)}$, then the lemma asserts that $10^{\lambda(b)} - 1$ divides $10^{k\lambda(b)} - 1$. We choose for k the particular value occurring in (3), and then we have

$$10^{k\lambda(b)} - 1 = 10^{\varphi(b)} - 1.$$

Therefore $10^{\lambda(b)} - 1$ is a divisor of $10^{\varphi(b)} - 1$. But according to § 6, b divides $10^{\lambda(b)} - 1$, and therefore b is a divisor of $10^{\varphi(b)} - 1$. We state this in the theorem:

If b is prime to 10, then $10^{\varphi(b)} - 1$ is divisible by b.

This theorem no longer has anything to do with decimal fractions, since $\varphi(b)$ has a completely independent meaning. Equation (2) gives us the special case:

If p is a prime that does not divide 10, then $10^{p-1} - 1$ is divisible by p.

The presence of the number 10 in these theorems is not essential. It occurs only because of the fact that our ordinary system of writing numbers is based on the number 10. If we think of using a number system based on any other number g, we can talk about "g-adic" instead of decimal fractions. Our whole discussion can be repeated without change, and we obtain the general theorems:

Theorem II. If b is prime to g, then $g^{\varphi(b)} - 1$ is divisible by b.

Theorem III. If p is a prime that does not divide g, then $g^{p-1} - 1$ is divisible by p.

Here we have found a theorem that extends far beyond the special topic of decimal fractions. It is a fundamental theorem of the theory of numbers. The special case III is called Fermat's theorem after its discoverer; theorem II is Euler's generalization of Fermat's theorem.

A few examples will illustrate these theorems:

$p = 5$:

$$2^{5-1} - 1 = 15 = 3 \cdot 5,$$
$$3^{5-1} - 1 = 80 = 16 \cdot 5,$$
$$4^{5-1} - 1 = 255 = 51 \cdot 5,$$

$p = 7$

$$2^{7-1} - 1 = \qquad 63 = \qquad 9 \cdot 7,$$
$$3^{7-1} - 1 = \qquad 728 = \qquad 104 \cdot 7,$$
$$5^{7-1} - 1 = \quad 15624 = \quad 2232 \cdot 7,$$
$$10^{7-1} - 1 = 999999 = 142857 \cdot 7,$$

$b = 6, \; \varphi(6) = 2:$

$$5^2 - 1 = 24 = 4 \cdot 6,$$
$$7^2 - 1 = 48 = 8 \cdot 6,$$

$b = 9, \; \varphi(9) = 6:$

$$2^6 - 1 = \qquad 63 = \qquad 7 \cdot 9,$$
$$4^6 - 1 = \quad 4095 = \quad 455 \cdot 9,$$
$$5^6 - 1 = 15624 = 1736 \cdot 9,$$

$b = 10, \; \varphi(10) = 4:$

$$3^4 - 1 = \qquad 80 = \qquad 8 \cdot 10,$$
$$7^4 - 1 = 2400 = 240 \cdot 10,$$
$$9^4 - 1 = 6560 = 656 \cdot 10.$$

11. After this digression, we now turn back to the decimal fractions. We have seen that the development of a reduced fraction $\dfrac{a}{b}$ with denominator b prime to 10 leads to a decimal fraction whose period begins immediately after the decimal point.

If such a decimal fraction is given, we would like to be able to find the common fraction from which it came. Let the decimal fraction have the period P of length λ. We will think of P as an ordinary number of λ digits. For example, for $\dfrac{1}{7} = 0.\overline{142857} \cdots$, we have $P = 142857$ (one hundred forty-two thousand eight hundred fifty-seven). If the decimal fraction is the development of $\dfrac{a}{b}$, then the remainder a will again appear after λ steps of the division of a by b. The part of the quotient obtained up to this appearance of a is exactly the period. Therefore $a \cdot 10^\lambda - a$ is divisible by b, and we have

$$a(10^\lambda - 1) = bP,$$

or

(4) $$\frac{a}{b} = \frac{P}{10^\lambda - 1}.$$

The fraction on the right will not usually be in reduced form, but on reduction it will give us the desired fraction $\frac{a}{b}$. Since 2 and 5 divide 10, they do *not* divide $10^\lambda - 1$, and they cannot divide b, which is a divisor of $10^\lambda - 1$. This shows that every periodic decimal fraction whose period begins immediately after the decimal point must have come from a common fraction $\frac{a}{b}$ in which the denominator is prime to 10. We already know that the converse is true. A decimal fraction in which the period begins immediately after the decimal point is called a "purely periodic" decimal fraction.

12. We have been considering only purely periodic decimal fractions. However, the examples under III in § 1 are not purely periodic. Such decimal fractions, in which one or more digits occur between the decimal point and the period, are called "mixed periodic." Since *finite* decimal fractions belong to fractions whose denominators have only the prime factors 2 and 5, and *purely periodic* ones belong to fractions whose denominators contain neither 2 nor 5, all that is left for the mixed periodic decimals is to belong to fractions whose denominators have a factor in common with 10 and a factor that is prime to 10. We merely mention this third case; it has nothing to offer that would repay our studying it.

13. After this discussion of the fundamental properties of periodic decimal fractions, we shall conclude this chapter with a consideration of a property which is more amusing than significant. The period of $\frac{1}{7}$ consists of the 6 digits 142857. We split them in half and add the numbers so formed:

$$142 + 857 = 999.$$

The period of $\frac{1}{17}$ is 0588235294117647 which, when split and added, gives

$$05882352 + 94117647 = 99999999.$$

For the period of $\frac{1}{11}$, which is 09, we have:

$$0 + 9 = 9.$$

We will show that the sum of the two halves of the period will always turn out this way when the period belongs to a fraction $\frac{a}{p}$ whose denominator is a prime, provided that the period has an even number of digits.

If the length λ of the period is even, we can write $\lambda = 2l$. Also, if the two halves of the period P are A and B, then P is a number of λ digits, while A and B have l digits each. Remembering the significance of the position of a digit in a number, we have

$$P = A \cdot 10^l + B.$$

Now we know that the fraction $\dfrac{a}{p}$ can be found from the period P by means of (4):

$$(5) \qquad \frac{a}{p} = \frac{P}{10^\lambda - 1} = \frac{A \cdot 10^l + B}{10^\lambda - 1}.$$

Since $\lambda = 2l$, we have

$$(6) \qquad 10^\lambda - 1 = 10^{2l} - 1 = (10^l - 1)(10^l + 1).$$

Equation (5) shows that the denominator p can be extended to $10^\lambda - 1$. This means that p divides $10^\lambda - 1$, a fact that we could have found from § 6. If p divides $10^\lambda - 1 = (10^l - 1)(10^l + 1)$, it must divide at least one of the two factors, because it is a prime. Now p cannot divide $10^l - 1$, because l is less than λ and λ is the *smallest* number for which $10^\lambda - 1$ is divisible by p (theorem I, § 6). Consequently p must divide $10^\lambda + 1$. From (5) and (6) we have

$$\frac{a}{p} = \frac{A \cdot 10^l + B}{(10^l - 1)(10^l + 1)},$$

and this can be rewritten as

$$\frac{a(10^l + 1)}{p} = \frac{A \cdot 10^l + B}{10^l - 1}.$$

The left side is a whole number because p divides $10^l + 1$. Therefore the right side is also a whole number. Now we also have

$$\frac{A \cdot 10^l + B}{10^l - 1} = \frac{A(10^l - 1) + A + B}{10^l - 1} = A + \frac{A + B}{10^l - 1},$$

and, since A is a whole number, so is

$$(7) \qquad \frac{A + B}{10^l - 1} = h.$$

Now A consists of l digits and is greatest when all these digits are 9's. The number consisting of l digits 9 is $10^l - 1$, and therefore we have $A \leqq 10^l - 1$. In the same way we find $B \leqq 10^l - 1$, and hence

$$A + B \leqq 2 \cdot (10^l - 1).$$

The equal sign cannot hold in this inequality, for $A+B=2 \cdot (10^l-1)$ implies that both A and B have the value $10^l - 1$. But then A and B would contain only 9's and P would be a sequence of $2l$ equal digits, 9. This is an absurdity, since the period would then consist of the *one* digit 9, while we have assumed that it has an even number of digits. Therefore we have

(8) $$A + B < 2 \cdot (10^l - 1).$$

Now (7) can be written as

(9) $$A + B = h(10^l - 1),$$

where h is a positive whole number. According to (8), h is less than 2, so it must be 1. Then (9) becomes

$$A + B = 10^l - 1,$$

that is, $A + B$ is the number of l digits $99 \cdots 9$.

24. A Characteristic Property of the Circle

When it rains, the ground is wet; when the ground is wet, it is not necessarily raining. This example is frequently used to elucidate the difference between a theorem and its converse. Clear as it may be in this formulation, it is very badly confused in ordinary life. Intelligent persons to whom the difference is crystal clear when it is brought to their attention are prone to mix it up unconsciously in ordinary intercourse. A political orator can often take a statement of his opponent and make it sound ridiculous by stating it in its converse form, without the trick's being noticed by his listeners. Every mathematician knows that in teaching he must systematically educate the beginning student to avoid committing an error by the unconscious use of an unproved converse.

But the conscious passage from a theorem to its converse is a useful and fruitful principle for the researcher in mathematics. This and the following chapter, which are otherwise independent, will show how this principle serves to lead from known theorems to new theorems and whole new concepts.

We begin with a simple example of a mathematical theorem and its converse. The theorem that all angles inscribed in a circle and subtended by the same chord are equal [1] (Fig. 84), is familiar from elementary geometry. Now the main point is that the converse of this theorem is also true: The locus of all points from which a given segment *AB* subtends equal angles is a circle. In elementary geometry the importance of this theorem and its converse is not brought out clearly. Since both the theorem and converse are true, this property of the inscribed angle is a characteristic of the circle. It could be used to replace the ordinary definition, that a circle is the locus of all points that are at the same distance from a fixed point. In fact, all of the most interesting theorems concerning circles are proved only after this theorem, and they depend on it more directly than on the definition of the circle.

Following these preliminary remarks, we turn to another property

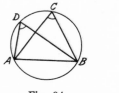

Fig. 84

Fig. 85

of the circle and show that it is a characteristic one. In elementary geometry an angle is defined (for good reasons, to be sure) as the amount of turning between two *straight* lines. However, there is no reason for us not to consider the amount of turning between two curves. If desired, it may be defined as the angle between the tangents to the two curves at the vertex (Fig. 85).

The circle has the obvious property that the chord joining any two points on it meets the circle at the same angle at the two points. These angles are angles between a line and a curve. We now form the converse and ask: *If a curve has the property that every chord joining every two points on it meets the curve at the same angle at the two points, is the curve always a circle, or are there other curves with this same property?* (Fig. 86). We shall show that it is always a circle, and hence that this is a characteristic property of the circle.

Let us suppose that we have a curve with this property, and

[1] If *D* lies on the lower arc *AB*, then the angle is to be measured between the extension of the ray *AD* through *D*, and the ray *DB* (otherwise supplementary angles must be considered).

that A, B, C are any three points on the curve (Fig. 87). We draw the three chords joining these points and the three tangents to the

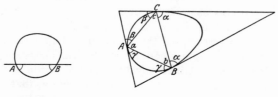

Fig. 86   Fig. 87

curve at these points. Then, in view of the assumed property, the angles designated by the same letters in the figure are equal.

The three angles at A form a straight angle, and the same is true of the angles at B and C. Therefore we have

$$a + \beta + \gamma = 2R,$$
$$\alpha + b + \gamma = 2R,$$
$$\alpha + \beta + c = 2R,$$

where R represents a right angle. Adding these equations, we find

$$(a + b + c) + 2(\alpha + \beta + \gamma) = 6R.$$

Now, since the sum of the three angles of a triangle is $2R$, we have

$$a + b + c = 2R,$$

and therefore

$$2(\alpha + \beta + \gamma) = 4R,$$
$$\alpha + \beta + \gamma = 2R.$$

Comparing this with

$$a + \beta + \gamma = 2R,$$

we obtain $a = \alpha$. In the same way we find $b = \beta$, $c = \gamma$.

Now let D be any other point on our curve. We can draw the triangle ABD and treat it exactly as we did the triangle ABC. Since the points A and B are the same in both triangles, the chord AB and the tangents at A and B are also the same. Because of this, the angle γ is the same in both figures. Since the angle d at D is equal to γ, and $\gamma = c$, we have $d = c$ for every point on the curve. Therefore the chord AB subtends equal angles at all points D on the curve. By the converse to the theorem on inscribed angles, this implies that D lies on the circle determined by the points A, B, C. Since D was any point on the curve, the curve must be a circle.

In contrast to this result, the next chapter will take up a property of the circle which is not characteristic. We shall find that a whole series of curves has this property, and we will see how the principle of the converse can lead to new concepts, in this case a remarkable class of curves.

25. Curves of Constant Breadth

1. A circle is defined as the curve all of whose points lie at a given distance from a fixed point, the center. The wheel is a direct practical application of this property of the circle. The hub of the wheel is held at a fixed height above the ground by the spokes of equal length, thus maintaining a smooth horizontal motion. In moving very heavy loads, the wheel and axle is sometimes not sufficiently strong. In this case one often resorts to the more primitive use of rollers. The load is merely rolled along over cylindrical rollers (Fig. 88) which are continually placed under the front. The load moves horizontally over these cylinders, whose cross sections are circles.

Obviously a wheel must be made in the form of a circle with the hub at the center, since any other form will produce an up-and-down motion. However, strange to say, it is not necessary that rollers have a circular cross section in order to perform their services properly. For rollers the *center* is no longer important. The property of the circle that allows it to be used for rollers is that every pair of parallel tangents is always at the same distance apart, no matter how the circle is turned. The circle has the same width

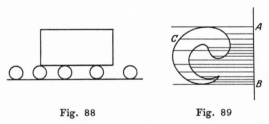

Fig. 88   Fig. 89

in every direction; it is what is called a "curve of constant breadth". One might expect this property of the circle to be completely characteristic, as were the properties discussed in the last chapter.

But surprisingly, there are other curves with this same property. Indeed, there is a great multiplicity of curves of constant breadth which are not circles.

2. If we wish to determine the breadth of some curve C in a given direction, we can project each point of the curve perpendicularly on a line parallel to that direction (Fig. 89). The projection will fill up some segment AB of the line, and the length of this segment will be the breadth of the curve in the given direction.

The two lines of projection that are perpendicular to AB at A and B have at least one point in common with the curve C, while the entire curve lies on only one side of the line. We shall call a line with this property a 'supporting line' of the curve.

A closed curve has exactly two supporting lines in each direction. They can be found by the method of Fig. 89, or we can draw two parallel lines in the given direction, containing the curve between them, and then slide them together until they just touch the curve (Fig. 90).

A supporting line is not the same as a tangent line. In Fig. 91a the line t is a tangent at T, but is not a supporting line. In Fig. 91b s is a supporting line but is not tangent to the curve.

For a curve of constant breadth, the distance between every pair of supporting lines is a fixed amount b. If we draw two pairs of

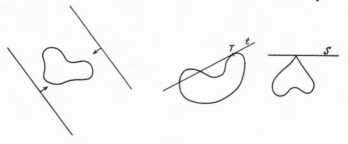

Fig. 90   Fig. 91a   Fig. 91b

supporting lines to such a curve, the parallelogram which they form will be a rhombus (Fig. 92). If the pairs of supporting lines are perpendicular, the rhombus is a square of side b. Therefore all squares circumscribed about a curve of constant breadth are congruent. This can be nicely illustrated by cutting a piece of heavy cardboard into the shape of a figure of constant breadth and cutting a square hole in another piece of cardboard. If the square has sides equal to the breadth of the curve, then it will fit the curve no matter what the direction in which it is turned. The curved figure can be turned freely inside the square without ever having any room to spare. Both this and its converse are true: A curve of constant breadth can be rotated inside a square without any space to spare,

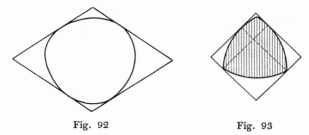

Fig. 92   Fig. 93

and a curve that can be rotated inside a square is a curve of constant breadth.

3.   The simplest curve of constant breadth that is not a circle is the curvilinear triangle pictured in Fig. 93. The three sides are equal arcs of circles, and the center of each arc is the opposite corner. The three arcs have equal radii which are equal to the constant breadth b of the curve. Of any two parallel supporting lines, one must touch at a corner and the other be tangent to the opposite side, or else both must touch at corners. In the first case, the distance between the supporting lines is clearly b. In the second case, each supporting line is tangent to the arc opposite the other corner, and their distance is again b.

This curvilinear triangle was first discovered, in the sense of a curve of constant breadth, by the technologist Reuleaux. He proved kinematically that this curve can be rotated inside a square without any space to spare. We have just seen that this property is characteristic of the curves of constant breadth.

4.   The principle used to construct the Reuleaux triangle can be extended to figures with more sides. The essential idea is to draw a series of arcs of equal radii in such a way that the center of each arc is the opposite corner. We can start with any point B for the first corner and draw an arc with radius b and B as center. On this arc we choose two points A and C to be new corners. The arc of radius b with center C goes through B, since $BC = b$ by the previous construction. On this arc we choose another corner D. The arc of radius b, with D as center, goes through C. If we wish to end this process, we choose the corner E on this arc so that it is

also on an arc of radius b with center A. That is, E is the intersection of these two arcs. Finally, we join A and D by an arc with

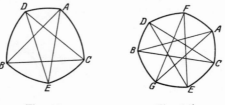

Fig. 94a   Fig. 94b

center E and obtain a curvilinear pentagon $ADBEC$ of constant breadth (Fig. 94a). Curvilinear polygons of more sides can be constructed in the same way by delaying the step at which the curve is closed. Fig. 94b is such a polygon of 7 sides. Since each corner is opposite a side which is an arc of radius b having the corner as its center, it immediately follows that this construction produces a curve of constant breadth b. For a later purpose we have joined each corner with the two ends of the opposite arc by means of radii. These radii form a self-intersecting polygon, all of whose sides are equal. The angle formed by each pair of radii through a corner is the central angle of the opposite arc.

All the curvilinear polygons constructed by this method will have an *odd* number of sides. To see this, we mark a corner and its opposite side. If we now pass around the curve starting at the marked corner, we will first pass a side, then a corner, and so on alternately until we pass a corner just before reaching the marked side. In all, we will have passed the same number, say n, of sides and corners in going from the marked corner to the marked side. Now if we start at the marked corner again and pass around the curve in the *opposite* direction, we will again pass n sides and n corners before reaching the marked side, since opposite each corner on the first path there is a side on the second path, and opposite each side there is a corner. Counting in the marked parts, there are then $2n + 1$ corners and the same number of sides.

5. The curves that we have constructed all have corners, that is, points where two sides meet at an angle. However, we can use these curves to obtain new curves of constant breadth that do not have any corners. Starting with one of our curves, we draw a curve parallel to it and outside it at a fixed distance d (Figs. 95a, b, c). This is easily done with the aid of the diagonal polygons

which we have drawn inside the curves. We merely replace each arc of the original curve by an arc having the same center, but

Fig. 95a   Fig. 95b   Fig. 95c

with the radius increased by d. The corners of the original figure are considered as arcs of radius 0, so they are replaced by arcs of radius d. The resulting figure is made up of an odd number of arcs of one radius and the same number of arcs of another radius. The arcs pair off, an arc of one radius being paired with one of the other radius having the same center (a corner of the original curve).

The same principle can be used to construct curvilinear polygons of constant breadth with arcs having radii of more than two different sizes. The opposite arcs are arranged so that they have the same center and so that their central angles form vertical angles (Fig. 96).

These methods allow us to construct an unlimited number of curves of constant breadth. However, these curves all have the

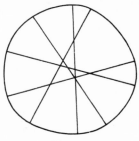

Fig. 96

special feature that they are formed of a number of circular arcs. In order to prevent a misunderstanding, we wish to emphasize that there are curves of constant breadth for which no part of the curve, no matter how small, is a circular arc.

6. Now that we have seen some examples of curves of constant breadth, we shall consider some of their general properties. In all of our examples the curves are convex curves, that is, curves which

have only two points in common with any line that cuts them. In order to simplify the discussion, we shall restrict our considerations to convex curves, and we shall always mean such curves even if we don't explicitly state that they are convex. As a matter of fact, this is really no restriction at all. It can be proved that every curve of constant breadth is convex. However, the proof would carry us too far afield, and we can avoid it by making the restriction.

Carefully defined, a convex curve is the boundary of a convex region. A convex region is characterized by the property that every two points of it can be joined by a straight line segment that is entirely in the region. Examples of convex regions are: a square, a circle, a triangle, an ellipse, and all the figures of constant breadth that we have mentioned. It is clear that a supporting line of a convex region will either have just one point or a whole segment in common with the boundary of the region. However, we shall prove the theorem:

Theorem I. A curve of constant breadth has just one point in common with each of its supporting lines.

Before proving this we make a simple observation:

Theorem II. The distance between any two points on a curve of constant breadth b is at most equal to b.

For if P and Q are two points on the curve (Fig. 97), then the two supporting lines perpendicular to the segment PQ must contain PQ between them. Therefore the distance between these lines is at least as large as the distance PQ. Since the distance between the supporting lines is b, the result is proved.

Turning to the proof of theorem I, we assume that it is false, that

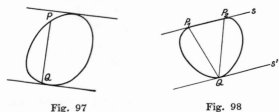

Fig. 97   Fig. 98

two points P_1 and P_2 of the curve lie on the supporting line s (Fig. 98). We draw the supporting line s' parallel to s on the other side of the curve and let Q be a point of contact of s' and the curve. The distance between s and s' is again b.

The segments P_1Q and P_2Q cannot both be perpendicular to s, since the triangle P_1QP_2 cannot have two right angles. Consequently one of the segments is longer than b, but this contradicts theorem II. Therefore the assumption of two points of the curve on the supporting line is disproved and we have theorem I.

If we again use the fact that the perpendicular line joining two supporting lines has length b while any other joining line is longer, we immediately obtain the theorem:

Theorem III. If a line joins the two points of contact of two parallel supporting lines of a curve of constant breadth, then it is perpendicular to the supporting lines.

7. If we draw a circle of radius b about any point of the curve of constant breadth b as center, then, by II, the whole curve will be enclosed by the circle. We shall show that the curve cannot lie wholly in the interior of the circle, but that it must have at least one point on the circumference.

Let P be any point on the curve C of constant breadth b. With P as center we draw a circle K_1 (Fig. 99) which is large enough to enclose C but small enough to have a point Q of C on its circumference. The radius r of K_1 is at most equal to b, since the circle K

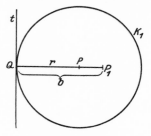

Fig. 99

of radius b and center P encloses C. Therefore K_1 is inside or at most identical with K.

The tangent t to the circle K_1 at Q goes through the point Q of C. Furthermore, C is enclosed by K_1 so it lies all on one side of t. Therefore t is a supporting line of C. The supporting line s, parallel to t on the other side of C, is at the distance b from t because C is of constant breadth b. According to theorem III, the point of contact P_1 of s is on the perpendicular to t through Q. If $r = b$, then P_1 falls on P; if $r < b$, then P lies between Q and P_1. But the latter is impossible. The three points Q, P, P_1 would belong to C and would lie on a line. Now a convex curve can be cut by a

169

line in only two points. A line could have more than two points in common with a convex curve only if it were a supporting line. But we know, according to I, that a supporting line has only one point in common with a curve of constant breadth. Therefore P_1 must fall on P, and we have $r = b$.

In this proof P was an arbitrary point of C, and we have constructed the supporting line s of C through this point. Therefore we have also proved the result:

Theorem IV. There is at least one supporting line through every point of a curve of constant breadth.

A curve of constant breadth may have points at which there is more than one supporting line. Such points are called corners. In our earlier examples there were many curves of constant breadth having corners. At a corner, all the lines that lie in the angle formed by two supporting lines are clearly supporting lines themselves (Fig. 100). Therefore a convex curve has a whole *bundle* of supporting lines at each corner. Among these supporting lines there are two extreme ones that bound the bundle.

If P is an arbitrary point of C, then by theorem IV we can draw s, the (or a) supporting line of C, at this point. We draw the per-

Fig. 100

pendicular to s through P. It will cut C in an opposite point Q, and PQ will have the length b. The circle of radius b with center Q will then enclose C and will have s as a tangent. This can be put in the form of the following theorem:

Theorem V. Through every point P of a curve of constant breadth, a circle of radius b can be drawn that encloses the curve and that is tangent, at P, to the supporting line of the curve, or to a predetermined supporting line if there are more than one.

8. The following theorem also relates to curves of constant breadth and circles:

Theorem VI. If a circle has three (or more) points in common with a curve of constant breadth b, then the length of the radius of the circle is at most b.

The Reuleaux triangle shows that such a circle may have a radius equal to b. If any one of the three arcs is extended to a full circle, this circle has radius b and has infinitely many points in common with the curve.

In order to prove theorem VI, we suppose that the circle k has the points P, Q, R in common with the curve C of constant breadth b. Of the three angles of the triangle PQR, there is at least one that is not exceeded by the other two, be it larger than both the others, equal to one and larger than the third, or, perhaps equal to both the others. We can suppose that this angle lies at P, and we shall call it α. Through P we now draw the (or a) supporting line of the curve of constant breadth. Then we draw the circle K of radius b, tangent to the supporting line at P and enclosing C. The points Q and R will lie inside or on the circumference of K. If both Q and R lie on the circumference, then K and k are identical, since there is only *one* circle that passes through three points P, Q, R. In this case there is nothing more to prove.

Otherwise we extend PQ and PR to their intersections, Q' and R', with K (Fig. 101). We now wish to prove that $Q'R'$ is longer than QR.

If Q and Q' happen to be the same point, then R and R' are different, since the case in which both Q and R lie on K has been settled. Here the triangles of Fig. 101 are related, as is shown in Fig. 102a. The angle $QRR' = \delta$ is an exterior angle of the triangle

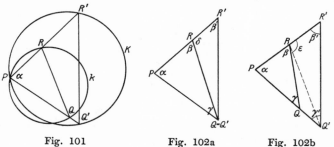

Fig. 101   Fig. 102a   Fig. 102b

PQR. Now, by a theorem of elementary geometry, an exterior angle is greater then either of the two angles of the triangle which are not adjacent to it. In our case we have $\delta > \alpha$. Also, since β is an exterior angle of the triangle QRR', we have $\beta > \beta'$. Because we chose $\alpha \geqq \beta$ in the triangle PQR, we have $\delta > \alpha \geqq \beta > \beta'$, or $\delta > \beta'$. Therefore QR' is a side of the triangle QRR' which is opposite the angle δ, while QR is opposite the smaller angle β'. Then, by a well-known theorem of geometry, we have $QR' > QR$.

If Q differs from Q' and R differs from R', the triangles are related as in Fig. 102b. Because of the theorem concerning the sum of the angles in a triangle, we have $\beta + \gamma = 180° - \alpha = \beta' + \gamma'$. It is therefore impossible to have $\beta' > \beta$ and $\gamma' > \gamma$ at the same time. Let us suppose that it is the first of these inequalities which does not hold. We then have $\beta' \leqq \beta$. In the quadrilateral $QQ'R'R$ we draw the diagonal $Q'R$, the diagonal which does not divide the angle β'. (In the case $\gamma' \leqq \gamma$, we would draw QR'.) Designating the angle $Q'RR'$ by ε, we see that it is an exterior angle of the triangle $PQ'R$, and we have $\varepsilon > \alpha$. Furthermore, since we already have $\alpha \geqq \beta$ and $\beta \geqq \beta'$, we finally get $\varepsilon > \beta'$. Then, in the triangle $Q'R'R$, the side $Q'R'$ is opposite an angle greater than the angle opposite $Q'R$, and we have $Q'R' > Q'R$. Since our earlier argument can be used to show that we also have $Q'R > QR$, we finally obtain the desired result $Q'R' > QR$.

We have now obtained $Q'R' > QR$ for every case in which the circles k and K (Fig. 101) are different. Now α, the angle inscribed in the circle k and subtended by the chord QR, is also subtended in the circle K by the chord $Q'R'$. Therefore the chords QR and $Q'R'$ belong to the same central angle 2α in the circles k and K respectively. If we bring these central angles together, we obtain Fig. 103. Here we

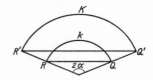

Fig. 103

recognize at once that the larger chord belongs to the larger circle, and therefore we see that the radius b of K is larger than the radius of k. This concludes the proof of theorem VI.

9. The simplest curve of constant breadth which is not a circle, the Reuleaux triangle, possesses corners. The following theorem shows that the Reuleaux triangle is outstanding among the curves of constant breadth because of its corners.

Theorem VII. A corner of a curve of constant breadth cannot be more pointed than 120°. The only curve of constant breadth that has a corner of 120° is the Reuleaux triangle, which has three such corners.

We measure the angle of a corner by means of the two extreme supporting lines of the bundle of supporting lines at the corner. If a corner Q has the angle ϑ, then the bundle of supporting lines occupies an angle of $180° - \vartheta$ (Fig. 104). The perpendiculars at Q to all these supporting lines form another bundle that occupies an angle of $180° - \vartheta$, the angle $P_1 Q P_2$. From theorem III we see that each of these perpendiculars crosses the curve at a point whose distance from Q is b .

Therefore the part of the curve opposite the corner Q is a circular arc of radius b and central angle $180° - \vartheta$. According to theorem II,

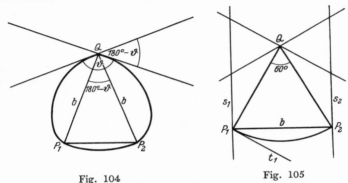

Fig. 104   Fig. 105

the length of the chord $P_1 P_2$ cannot exceed the width b . Then the isosceles triangle $Q P_1 P_2$ has legs of length b, and its base cannot exceed b in length. Therefore the angle $P_1 Q P_2$ is at most 60°. We have already seen that this angle $P_1 Q P_2$ is $180° - \vartheta$, so we have $180° - \vartheta \leqq 60°$, and hence $\vartheta \geqq 120°$. Since ϑ was the angle at an arbitrary corner, the first part of the theorem is proved.

Now if the corner angle ϑ is 120°, the angle $P_1 Q P_2$ is 60° and the isosceles triangle $P_1 Q P_2$ is equilateral (Fig. 105). Then $P_1 P_2$ has the length b. Since this length is equal to the breadth of the curve, the two supporting lines perpendicular to $P_1 P_2$ must pass through P_1 and P_2. From this we can see that P_1 and P_2 must also be corners of the curve. The part of the curve between P_1 and P_2 is an arc of a circle, as we have already seen. Not only is s_1 a supporting line at P_1, but so is the tangent t_1 to the circular arc at P_1. The inner angle between these two lines is easily seen to be 120°. Consequently s_1 and t_1 must be the extreme supporting lines of the bundle through P_1, since no corner can have an angle less than 120°. The corner at P_1 (and similarly at P_2) therefore has exactly the angle 120°. Then P_1 and P_2 have just the same properties as Q. Opposite each of them is a circular arc of radius b and central angle 60°. But this gives us exactly the Reuleaux triangle, and therefore the second part of theorem VII is proved.

10. In this chapter we first obtained some special curves of constant breadth by seeing how to construct them. Then we proved the general properties given by theorems I to VII. These properties hold for all convex curves of constant breadth, but they do not say anything about the existence of curves of constant breadth. We will now give a perfectly general construction that will yield every curve of constant breadth. This will give us a complete view of all possible curves of constant breadth. The property of our curves shown in theorem V is an especially important one. Curves of constant breadth are characterized by this property to the extent that one may arbitrarily choose one-half of such a curve between two opposite points, so long as it satisfies the conditions of theorem V. More accurately stated, we assert:

Theorem VIII. If a convex arc[1] Γ *has a chord of length b, if the entire arc lies between the two perpendiculars to the chord at its ends, and if it has the property of being enclosed by every circle of radius b tangent to a supporting line at its point of contact and lying on the same side of the line as the arc, then the curve can be extended to form a curve of constant breadth b.*

11. In proving this theorem it will be convenient to use the idea of *regions* of constant breadth. Since every region of constant breadth is bounded by a *curve* of constant breadth, we need only show the existence if a suitable region.

Before starting the proof we must make a remark about 'intersections' of regions. If a number of regions are given, then the part of the regions that is common to all of them is called their intersection. For example, the intersection of the two circles in Fig. 106 is the shaded region.

The intersection of an arbitrary set of convex regions is itself convex.

To prove this we must show that every two points of the intersection may be joined by a line segment that lies entirely within the intersection. But this is obvious. For if two points P and Q lie in the intersection, they lie in every region of the set. Then, because each region of the set is convex, the segment PQ is in all the regions. Since it is in all the regions, the segment PQ must also be in the intersection.

---

[1] That is, an arc which, together with its chord, bounds a convex region.

In this proof it is quite immaterial whether the set contains finitely or infinitely many convex regions.

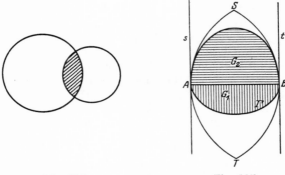

Fig. 106   Fig. 107

12. We now suppose that the arc Γ (Fig. 107) has the properties required by theorem VIII: the chord AB has the length b. The arc, together with its chord, bounds the convex region G_1. The perpendiculars to AB at A and B are supporting lines of G_1. Every circle of radius b that is tangent to a supporting line of Γ at its point of contact encloses Γ.

To G_1 we add a new region ABS bounded by the chord AB, the circular arc AS with center B, and the circular arc BS with center A. We shall call this region G_2. The two convex regions G_1 and G_2 together form a convex region G which is bounded by Γ and the arcs AS and BS.

We now consider the totality of all circles of radius b whose centers lie on Γ. The region G and this infinite set of circles have a convex intersection D (the shaded part of the figure). We will now demonstrate that this region D is a region of constant breadth having the arc Γ as part of its boundary.

If Γ belongs to D it must certainly be on the boundary, since it is already on the boundary of G. But Γ is in all the circles of radius b with centers on Γ. To demonstrate that this is true, we need to show that no two points of Γ can be further apart than the distance b. Now, by assumption, Γ lies in both the circles of radius b with centers A and B, and therefore it lies in the curved figure $SATBS$. Since Γ lies on one side of AB, it must lie in the region G_2', which is the mirror image of G_2 in AB. The distance between any two points of G_2' is clearly at most b, so this must be true in particular for any two points on Γ. Therefore Γ belongs to D and it is a part of the boundary of D. Since D is convex, it must contain every chord joining two points of Γ. For this reason G_1 must be a part of D.

No two points of D can be further apart than the distance b. Since D is a part of G and G is made up of the two regions G_1 and G_2, we have three cases to consider: If both points are in G_1, then, as we mentioned before, they cannot be further apart than the distance b. If both points are in G_2 the same is true. Finally, if one point P_1 is in G_1 and the other point P_2 is in G_2, we can join P_1 with P_2 and extend the line until it cuts Γ, say at P. The three points will lie on this line in the order PP_1P_2. The circle of radius b with center P contains all of D, so it contains these three points. Consequently P_1 and P_2 lie on a radius of length b, and hence the distance between them cannot exceed b.

The result we have just obtained shows that the region D cannot have a breadth greater than b in any direction. We must now show that it has the breadth b in every direction. In the direction AB, the breadth b was prescribed by the theorem. We consider any other direction and draw the two supporting lines of D perpendicular to this direction. One of the two, say s_1, will have a point of contact Q on Γ. At Q we draw a perpendicular of length b to s_1 and call its end point M. Now M belongs to D. To prove this we must show that M is in G, as well as in every circle of radius b with center on Γ. The latter requires that we show that the distance of M from each point of Γ is at most b. This follows from the fact that the circle of radius b with center M is tangent to s_1 at Q. According to the assumptions, this circle must enclose the arc Γ, and this shows that the distance, from M to any point of Γ is not greater than b. Since in particular the distances AM and BM are not more than b, the point M must lie in the figure $SATBS$. Furthermore, since M lies on the opposite side of AB from Γ, it also lies in G_2. Therefore it lies in G as well as in all the circles. Consequently M lies in the intersection D.

Since QM is perpendicular to s_1 and s_2, and Q and M belong to D, the distance between the supporting lines s_1 and s_2 must be at least as large as $QM = b$. The distance cannot be greater than b, since this would mean that the distance between the two points of contact was greater than b, and we have seen that this is impossible for two points of D. Therefore the distance between s_1 and s_2 is exactly b; the region D has breadth b in every arbitrary direction.

This theorem shows that the arc Γ, satisfying certain requirements, can be extended to form a curve C of constant breadth b. It can easily be seen that there is only one way in which the extension can be made, and that the curve C is therefore uniquely determined.

13.   In conclusion we mention, without proof, another remarkable property of these curves: All curves of constant breadth having the same breadth b have the same perimeter. This perimeter must obviously equal the circumference of a circle of diameter b. This fact can easily be verified for the examples constructed in §§ 4 and 5, making use of the similarity of circular arcs with the same central angle. However, the proof for general curves requires ideas and methods which are beyond the scope of this book. The proof can be started only after a very careful analysis of the concept of the length of a curve.

26.   The Indispensability of the Compass for the Constructions of Elementary Geometry

1.   The constructions of elementary geometry are all carried out with the aid of a straightedge and compass.   In fact, a distinguishing property of elementary geometry is the fact that the only implements allowed are the compass and straightedge.   But these two instruments are not entirely necessary.   There are many constructions in which one or the other can be dispensed with.   More than this, according to the investigations of Mascheroni and the recently found earlier work of Mohr, the straightedge can be dispensed with entirely. All constructions that are possible with a straightedge and compass can be made with a compass alone.   Since a line cannot be drawn without a straightedge, in these investigations a line is considered as being constructed as soon as two of its points are found. On the other hand, Jacob Steiner has found that all the constructions of elementary geometry can be made using only a straightedge, provided only that *a fixed circle and its center* have been drawn beforehand.   It is not difficult to prove that this fixed circle is indispensable.   We shall prove this by showing that a *fixed circle whose center is unknown* is not sufficient to allow all the constructions to be carried out with a straightedge alone.   Indeed, *two non-intersecting circles* with unknown centers will not suffice.   However, it is known that two intersecting circles without their centers, or three non-intersecting circles, are sufficient to replace the Steiner circle with center.

2. Since according to Steiner's result all the constructions of elementary geometry can be carried out as soon as we have a given circle and its center, what we must prove is that it is impossible, using only a straightedge, to construct the center of the circle if we have been given just the circle or two non-intersecting circles. The proofs of the impossibility of each of these two constructions will be made by showing that the assumption of a construction leads to an absurdity. These *indirect proofs* will make use of the principle of mapping.

If we prove that it is impossible to find the centers of two non-intersecting circles with the straightedge alone, then clearly the impossibility of finding the center in the case of one given circle will follow immediately. But the latter case is geometrically simpler, so we shall prove it first. Also it will serve to introduce the ideas that are used in both proofs.

3. Suppose we have some way of finding the center of a circle drawn on a piece of paper without using anything but a straightedge. Our construction will consist in drawing lines which may cut the circle or each other, as well as lines which join intersections that we have already found. We will have determined the center of the circle when we have found two lines whose intersection is the center. The whole figure will then be composed of the given circle and a number of lines, two of which intersect at the center of the circle.

We will now study a particular mapping of this whole figure. This mapping will carry the given circle into a circle, every straight line into a straight line, and the intersection of two lines into the intersection of the two corresponding lines in the map. There are obviously many such mappings. For example, any proportional magnification or contraction of the figure is such a mapping. But a proportional mapping of this sort will not serve our purpose. We will have to find a mapping that carries the circle into a circle and every line into a line, but that otherwise distorts the figure. In particular, we want the center of the circle to be mapped onto a point that is not the center of the image circle.

Once we have found the desired mapping, our proof will be practically complete. No matter how much the image differs from the original figure, the two are equivalent so far as the construction is concerned. Every step of the original proof, drawing a line, finding an intersection, joining two intersections, can be carried out step by step in the image. But the image of the center of the original circle is *not* the center of the image circle. Therefore the images of the lines that intersect at the center of the original circle do not intersect at the center of the image circle. Although the construction was carried out step for step in the image, it does not lead to the center of the circle in the image. This contradicts the assumption that the construction does determine the center of the circle, and hence shows that it is impossible, using a straightedge, to find the center of a given circle.

For the case of *two* circles the proof will be quite analogous.

4. Now we must find a mapping of the sort we have described. We shall obtain it by means of a projection in space. Outside of the plane E of the figure (Fig. 108) we mark a fixed point O and draw another plane E', the image or projection plane. Each ray through O and a point P on E, produced if necessary, will intersect

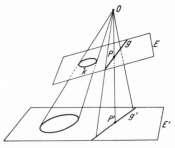

Fig. 108

E' at some point P'. Then P' is the image or projection of P. In the same way, a whole figure in E is projected point for point into a figure on E'. The projection can be thought of as the shadow cast on E' by the figure in E, if O represents a point-source of light. Under the projection every line g will project into a line g', since the totality of rays through O and the points of g forms a plane and this plane cuts E' in a line.

The projection of a circle will not generally be a circle. The totality of rays through O and the circumference of the circle k forms a cone. In general this will be an oblique cone. A cone is called a "right" cone if the line connecting its vertex to the center of its base is perpendicular to the base. An oblique cone is one that is not right. The projection plane E' cuts the cone in a conic section which is known not to be a circle in general. However, it is essential for our purposes that the circle project into a circle. There are two particular arrangements of the projection that will accomplish this.

The first, trivial, way is to place the planes E and E' parallel to each other. Then the mapping performed by the projection is a proportional mapping, a magnification or contraction according to whether E or E' lies nearer O. This mapping is useless for our purposes because it fails to distort the figure.

The second way to accomplish the mapping depends on a theorem of solid geometry. So as not to interrupt the course of our discussion, we shall postpone the proof of this theorem to § 8. The plane that is perpendicular to the base of the oblique cone, and that contains the vertex O and the center of the base of the cone, is a plane of symmetry of the oblique cone. Fig. 109 represents the intersection of the cone with this plane. Only the diameter K_1K_2 of the circular base appears in this figure. The plane of the base is

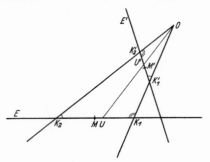

Fig. 109

perpendicular to the plane of the figure. Because of the position of the plane of the figure, the line OK_1 is the shortest of all the rays from O to the circumference of the base, while OK_2 is the longest of all these rays. Every plane parallel to the base plane E obviously intersects the cone in a circle. If the plane E' cuts the lines OK_1 and OK_2 at K_1' and K_2' in such a way that $\angle\, OK_1'K_2' = \angle\, OK_2K_1$, then the theorem that we shall prove in § 8 asserts that E' intersects the cone in a circle. Because we have taken $\angle\, OK_1'K_2' = \angle\, OK_2K_1$, we also have $\angle\, OK_2'K_1' = \angle\, OK_1K_2$, since the sum of the angles of a triangle is 180°. Clearly any plane parallel to E' will also cut the cone in a circle.

With E' chosen as we have just described, the projection from E to E' has all the properties that we require. We have only to verify that the midpoint M of K_1K_2 is *not* mapped on the midpoint M' of $K_1'K_2'$. In the first place, the bisector of the angle at O is the same line, whether we think of the angle as belonging to the triangle K_1OK_2 or to the triangle $K_1'OK_2'$. Now in any triangle the bisector of an angle divides the opposite side into segments proportional to the adjacent sides of the triangle. In the oblique cone we have $OK_2 > OK_1$, and therefore $K_2U > UK_1$ if U is the intersection of E and the bisector of the angle at O. Furthermore we have $\angle OK_1K_2 > \angle OK_2K_1$, since the larger angle is opposite the larger side in a triangle. Then we have $\angle OK_2'K_1' > \angle OK_1'K_2'$ and hence $OK_1' > OK_2'$, from which we get $U'K_1' > K_2'U'$. Since M is the midpoint of K_2K_1 and M' is the midpoint of $K_2'K_1'$, we now see that U and U' lie on opposite sides of the bisector of the angle at O. Being in this position, M' cannot be the image of M under our projection from O.

5. This completes our proof that the unknown center of a given circle cannot be found by means of the straightedge alone. Using Fig. 109, we can summarize the main points of the proof. A figure in the plane E, consisting of the circle K_1K_2 and certain lines, is projected from O onto the plane E'. Under this projection, lines go into lines and the circle K_1K_2 goes into the circle $K_1'K_2'$, but the center M of K_1K_2 does *not* go into the center M' of $K_1'K_2'$. Any construction based on lines that determine the center in E will not do so in E'. Therefore the construction under consideration is impossible.

6. The mapping is harder to obtain in the case of two given circles. Projection from a point O forms two oblique cones, and we must arrange it so that the projection plane E' cuts them both in circles.

We distinguish two cases. First, *one of the circles may lie inside the other.* We draw a figure (Fig. 110) whose plane is perpendicular to E and passes through the centers M and N of both circles. The circles are represented by their diameters K_1K_2 and L_1L_2 in the figure. Now if we can place the point O so that the angles K_1OK_2 and L_1OL_2 have the same bisector, then we can draw E' in such a way that $\angle L_2WO = \angle L_1'W'O$, using the notation of Fig. 110. Then all the angles designated by the same number or letter in the figure are easily seen to be equal. Therefore, according to the theorem of § 8, the plane E' cuts both cones in circles $K_1'K_2'$ and $L_1'L_2'$.

The remainder of the proof is the same as before: the projection carries the two circles into two circles and it carries lines into lines, but the centers of the circles are not carried into the centers of their projections. Any straight line construction in E determining the centers will fail in E', and in general no such construction is possible.

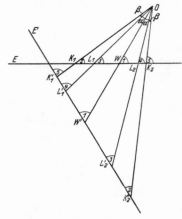

Fig. 110

We must still show that we can choose O in such a way that the angles K_1OK_2 and L_1OL_2 have the same bisector. For this to be true, we must have $\angle K_1OW = \angle K_2OW$ and $\angle L_2OW = \angle L_1OW$. Adding these, we see that we must have $\angle K_1OL_2 = \angle K_2OL_1$. We choose any arbitrary value δ for these angles and construct the isosceles triangles $K_1C_1L_2$ and $L_1C_2K_2$ with base angles $90° - \delta$ (Fig. 111). Using C_1 as center, we draw a circular arc through K_1 and L_2. Using C_2 as center, we draw a circular arc through L_1

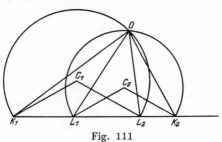

Fig. 111

and K_2. These arcs will intersect because their chords K_1L_2 and L_1K_2 overlap. We choose O as their intersection. Now we have

$$\angle K_1OL_2 = \tfrac{1}{2} \angle K_1C_1L_2 = \tfrac{1}{2}\left[180° - 2(90° - \delta)\right] = \delta,$$
$$\angle L_1OK_2 = \tfrac{1}{2} \angle L_1C_2K_2 = \tfrac{1}{2}\left[180° - 2(90° - \delta)\right] = \delta.$$

Consequently we have $\angle K_1 O L_2 = \angle L_1 O K_2$, from which we find $\angle K_1 O L_1 = \angle K_2 O L_2$. Therefore the bisector of $\angle L_1 O L_2$ is also a bisector of $\angle K_1 O K_2$, as was required.

7. In the second of our two cases *the two given circles lie outside of each other*. Here the two cones will be outside of each other, and it will be impossible for the angles $K_1 O K_2$ and $L_1 O L_2$ to have the same bisector for any position of O.

In this case we will have to use the other nappe of the cone (Fig. 112). The theorem of § 8 will apply to both cones if $\angle L_1 V O = \angle L_2' V' O$ and $\angle K_2 U O = \angle K_1' U O$ Then the triangle UVV' will

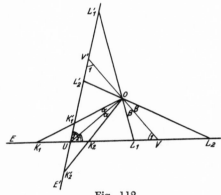

Fig. 112

be an isosceles triangle, and UO will bisect its vertex. Therefore UO is perpendicular to VV', and we have

$$\angle K_1 O L_1 = 90° + \alpha - \beta,$$
$$\angle K_2 O L_2 = 90° - \alpha + \beta,$$

and consequently

(1) $$\angle K_1 O L_1 + \angle K_2 O L_2 = 180°.$$

Therefore the point O must be chosen to make (1) true. Also, if (1) is true, then the bisectors of the angles $K_1 O K_2$ and $L_1 O L_2$ will actually be perpendicular, for we have

$$2 \angle UOV = 2 \angle UOK_2 + 2 \angle K_2 O L_1 + 2 \angle L_1 OV$$
$$= \angle K_1 OU + \angle UOK_2 + \angle K_2 O L_1 + \angle K_2 O L_1 + \angle L_1 OV + \angle VOL_2$$
$$= \angle K_1 O L_1 + \angle K_2 O L_2 = 180°,$$

and therefore

$$\angle UOV = 90°.$$

If this is the case we can place E', as shown in the figure, in such a position that the angles 2 are equal, and then the angles 1 will also be equal. Then E' cuts both cones in circles, the projection has the required properties, and the remainder of the proof is the same as before.

In order to find the position of the point O satisfying (1), we choose arbitrary values $\angle K_1OL_1 = \varphi$ and $\angle K_2OL_2 = \psi$, subject to the restriction $\varphi + \psi = 180°$. Then, following the method used at the end of § 6, we draw a circular arc in which the angle φ can be inscribed over the chord K_1L_1. Similarly, we draw an arc in which ψ can be inscribed over K_2L_2. Since K_1L_1 and K_2L_2 overlap, the circular arcs will intersect. Their intersection is the point O that we are looking for.

We have now proved that two non-intersecting circles without centers are not sufficient to make all elementary constructions possible with a straightedge alone.

8. There still remains the proof of the theorem postponed from § 4. This theorem asserts that *if the plane E* (Fig. 109) *intersects the oblique cone K_1OK_2 in a circle, then the plane E' will also intersect the cone in a circle if the angles $K_1'K_2'O$ and K_2K_1O are equal.* It must be remembered that the plane of the figure is the plane perpendicular to the base of the cone that is determined by the vertex O and the center M of the base. The condition $\angle K_1'K_2'O = \angle K_2K_1O$ can clearly be replaced by $\angle K_2'K_1'O = \angle K_1K_2O$ or by $\angle K_2'U'O = \angle K_1UO$.

The plane of Fig. 109 is a plane of symmetry of the cone. Our proof will depend on finding another plane of symmetry.

Fig. 113 shows the same cross section of the cone as is depicted in Fig. 109. We have drawn the circle circumscribing the triangle OK_1K_2 as well as the bisector MO of the angle at O. Since the angles K_1OM and K_2OM are equal, the arcs K_1M and K_2M which they intercept are equal. Therefore the perpendicular from the center C of the circle to the chord K_1K_2 intersects the circle at the point M. If we rotate the circle about the axis $M'CM$, it describes a sphere. The circumference of the base K_1K_2 and the vertex O of the cone lie on this sphere; the cone is inscribed in the sphere. Therefore the point M is equally distant from all points on the circle K_1K_2.

The line OM plays a special role in the study of oblique cones. It is called the *axis of the cone*. If a plane is passed through the axis, it cuts the cone in a triangle with one vertex at O and the other two, H_1 and H_2, on the base of the cone. This plane cuts the sphere in a circle that circumscribes the triangle OH_1H_2. The

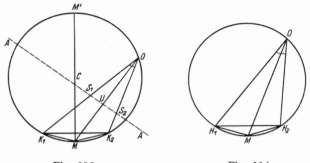

Fig. 113   Fig. 114

point M also lies on this circle, and we obtain a figure (Fig. 114) that is completely analogous to Fig. 113. Since the two chords H_1M and H_2M are equal because of the relation of M to the circular base of the cone, the axis OM is the bisector of the angle H_1OH_2. *The axis of an oblique cone has the property that every plane through it cuts the cone in a triangle in which the axis is a bisector of an angle.*

Now we cut the cone with a plane A perpendicular to the axis of the cone and to the plane of Fig. 113. The intersection of A and the plane of the figure is shown as the dotted line AA in Fig. 113. The intersection of A and the cone is represented in Fig. 115. The plane of this figure is the plane A, and the axis of the cone is perpendicular to it. The plane of Fig. 113 cuts the plane of Fig. 115 in

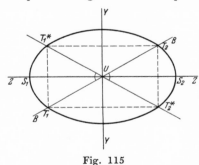

Fig. 115

the line ZZ. We consider an arbitrary plane through the axis. It cuts the plane of Fig. 115 in some line, say BB. The points T_1,

T_2 represent its intersection with the curve. The vertex O is not in the plane of Fig. 115 but it stands directly above the point U. In the triangle T_1OT_2, the axis UO is perpendicular to T_1T_2 and, by our above result, it is the bisector of angle T_1OT_2. Therefore we have $T_1U = T_2U$. Since B was any plane through the axis, this means that the point U is a center of symmetry of the intersection of the cone and the plane A.

The plane of Fig. 113 is a plane of symmetry of the cone, and its intersection with the plane of Fig. 115 is ZZ. Therefore Fig. 115 is symmetrical with respect to both the line ZZ and the point U. In other words, the mirror images T_1^* and T_2^* on the cone correspond to T_1 and T_2 on the cone.

The four points form a quadrilateral $T_1T_2^*T_2T_1^*$. Since T_1U and T_2U are equal, we have $T_1^*U = T_2^*U$ in the mirror image. Furthermore, $\angle \, T_1UT_1^* = \angle \, T_2UT_2^*$, since they are vertical angles. Then the triangles $T_1^*T_1U$ and $T_2^*T_2U$ are congruent, so we have $T_1T_1^* = T_2T_2^*$. The lines $T_1T_1^*$ and $T_2T_2^*$ are also parallel, since they are both perpendicular to ZZ. Therefore $T_1T_2^*T_2T_1^*$ is a parallelogram. Since the two diagonals T_1T_2 and $T_1^*T_2^*$ are equal, the parallelogram is a rectangle and U is its center. This shows that Fig. 115 also has the axis of symmetry YY.

Our result shows that the plane through the axis, that is perpendicular to the planes of Figs. 113 and 115, is a second plane of symmetry of the cone. It is only necessary to notice that this plane contains the line YY which is shown in Fig. 115 and which passes through U perpendicular to the plane of Fig. 113.

If we reflect the cone in this second plane of symmetry, its image coincides with the cone itself; this is just another way of stating the symmetry. However, the image of the circle K_1K_2 is another circle

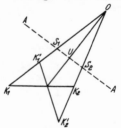

Fig. 116

$K_1'K_2'$, which must also lie on the cone (Fig. 116). This is the circle we have been looking for. The angle between the plane of this circle and the axis of the cone is in agreement with the statement of the theorem. Finally, we note that the circle $K_1 K_2$ cannot coincide with its image, since this could happen only if its plane were perpendicular to the axis of the cone. But this is in contradiction to the assumption that the cone is oblique.

27. A Property of the Number 30

Neither 10 nor 21 is a prime number. But $10 = 2 \cdot 5$ and $21 = 3 \cdot 7$ have no divisor that is common to them both. For this reason they are called "relatively prime" numbers. The numbers 6 and 10 are not relatively prime; they have the common divisor 2.

Of all the numbers from 1 to 9, the numbers 3, 7, 9 are relatively prime to 10. Although 9 is relatively prime to 10, it is not a prime number. In the case of 12 the situation is different. Of the numbers from 1 to 11, only 5, 7, 11 are relatively prime to 12, and they are all prime numbers. It can easily be seen that this property of 12 is shared by the numbers

$$3, \ 4, \ 6, \ 8, \ 12, \ 18, \ 24, \ 30.$$

Is 30 *the largest number that has this property, that all the numbers less than it and relatively prime to it are prime numbers?* This chapter will be devoted to show that this is so.

We start by seeing how we might set about looking for numbers with this property. From 4 on, every such number N must be divisible by 2. For if it were odd, 4 would be relatively prime to it, while 4 is not a prime number. In the same way, every such N from 9 on must be divisible by 3. Since it is already divisible by 2, it must be divisible by $2 \cdot 3$. Continuing this argument, we obtain the table:

From 4 on, N must be divisible by	$2 =$	2
From 9 on, N must be divisible by	$2 \cdot 3 =$	6
From 25 on, N must be divisible by	$2 \cdot 3 \cdot 5 =$	30
From 49 on, N must be divisible by	$2 \cdot 3 \cdot 5 \cdot 7 =$	210
From 121 on, N must be divisible by	$2 \cdot 3 \cdot 5 \cdot 7 \cdot 11 =$	2310.

Between 4 and 9, the only possible values for N are 4, 6, 8; between 9 and 25, only 12, 18, 24; between 25 and 49, only 30 (60, the next multiple of 30, is larger than 49). Between 49 and 121 there are no possibilities, since 210 is already larger than 121. Now we see that if the table continues in this way, that is if 13^2 is less than $2 \cdot 3 \cdot 5 \cdot 7 \cdot 11$, 17^2 is less than $2 \cdot 3 \cdot 5 \cdot 7 \cdot 11 \cdot 13$, and so on, then 30 is the largest number having the property we require. If we represent the successive prime numbers

$$2, \ 3, \ 5, \ 7, \ 11, \ 13, \ 17, \ \cdots,$$

by the symbols

$$p_1, \ p_2, \ p_3, \ p_4, \ p_5, \ p_6, \ p_7, \ \cdots,$$

as we did in Chapter 1, then we must show that

(1) $$p_{n+1}^2 < p_1 p_2 p_3 \cdots p_n,$$

is true for all n from 4 on.

Euclid's proof, which is reproduced in Chapter 1, shows that we have

$$p_{n+1} < p_1 p_2 \cdots p_n.$$

What we need is

$$p_{n+1} < \sqrt{p_1 p_2 p_3 \cdots p_n},$$

and this asserts more than Euclid's result. The inequality (1) is probably of as much interest as the original problem concerning the number 30. Since the original problem is solved as soon as we have (1), we shall concentrate on the proof of (1).

The inequality (1) asserts far less than is actually true. Not only is the next prime number after $p_5 = 11$ less than $\sqrt{2 \cdot 3 \cdot 5 \cdot 7 \cdot 11} = \sqrt{2310} = 48.06 \cdots$, it is only 13. The discrepancy becomes even greater as we go on. However, because of the tremendous irregularity of the primes, it is very difficult to obtain results that are valid for all primes. With the aid of extensive methods, Tschebyscheff proved that the next prime after p_n is less than $2p_n$, that is, $p_{n+1} < 2p_n$. This is much more than we need here, so the question arises whether our lesser assertion (1) cannot be proved, through the use of elementary methods.

H. Bonse, as a student, discovered an ingenious proof of (1). It not only avoids all the analytical methods and infinite processes that were used by Tschebyscheff, but it uses only the very simplest mathematical ideas.

1. The basic idea of the proof is similar to that of Euclid's proof, as given in Chapter 1. Instead of forming the expression

$$N = p_1 p_2 \cdots p_n + 1 \ \text{ or } \ M = p_1 p_2 \cdots p_n - 1$$

out of the first n prime numbers, we use only the first i prime numbers $p_1, \cdots, p_i$ and form the p_i expressions

$$M_1 = p_1 p_2 \cdots p_{i-1} \cdot 1 - 1,$$
$$M_2 = p_1 p_2 \cdots p_{i-1} \cdot 2 - 1,$$
$$M_3 = p_1 p_2 \cdots p_{i-1} \cdot 3 - 1,$$
$$M_4 = p_1 p_2 \cdots p_{i-1} \cdot 4 - 1,$$
$$\cdots\cdots\cdots\cdots\cdots\cdots$$
$$M_{p_i} = p_1 p_2 \cdots p_{i-1} \cdot p_i - 1.$$

As in the case of Euclid's expression, we can assert:

(a) *None of the expressions $M_1, \cdots, M_{p_i}$ is divisible by any of the prime numbers $p_1, p_2, \cdots, p_{i-1}$.*

(b) *At most one of these expressions is divisible by p_i.* For if two of them, say $p_1 \cdots p_{i-1} x - 1$ and $p_1 \cdots p_{i-1} y - 1$, were divisible by p_i, their difference $p_1 \cdots p_{i-1}(x - y)$ would also be divisible by p_i. Since p_i does not divide any of the first $i - 1$ factors, it would have to divide $(x - y)$. But x and y are among the numbers 1, 2, 3, 4, 5,$\cdots$, p_i, so their difference is at most $p_i - 1$, which is less than p_i. Therefore p_i cannot divide this difference, since a larger number cannot divide a smaller one. This same proof shows that we can also assert:

(c) *At most one of the expressions is divisible by p_{i+1}, at most one by $p_{i+2}, \cdots$, at most one by p_n.*

Now if there are fewer of the numbers $p_i, p_{i+1}, \cdots, p_n$ than there are of the expressions $M_1, \cdots, M_{p_i}$, in other words, if we have

(2) $$n - i + 1 < p_i,$$

then at least one of the expressions M is not divisible by any of the numbers $p_i, p_{i+1}, \cdots, p_n$. This important step in the proof follows directly from (b) and (c). If we call this particular expression M_h, then M_h is not divisible by any of the prime numbers $p_1, \cdots, p_n$, since (a) shows that it is not divisible by $p_1, \cdots, p_{i-1}$. The next step follows Euclid's proof. Either M_h is a prime number or it can be factored into prime factors. There is a prime number p which is either equal to M_h or divides M_h. Since M_h is not divisible by any of $p_1, \cdots, p_n$, this prime number p must be beyond p_n. The next prime number after p_n is p_{n+1}, so we have $p_{n+1} \leqq p$. Also, since p divides or is equal to M_h, we have $p \leqq M_h$. The largest of all the expressions M is M_{p_i}, so we can put these inequalities together to obtain

$$p_{n+1} \leqq M_{p_i} = p_1 \cdots p_{i-1} p_i - 1 < p_1 \cdots p_i.$$

We can summarize our results in the statement: *If* (2) *holds, then we have*

(3) $$p_{n+1} < p_1 \cdots p_i.$$

2. The result of § 1 is an improvement of Euclid's result,

$$p_{n+1} < p_1 \cdots p_n.$$

The number i is less than n, so the right side has been decreased. How much have we actually gained? The condition (2) does not allow us to choose an arbitrarily small value for i. We must keep i large enough so that the number of numbers $p_i, \cdots, p_n$, that is, $n - i + 1$, is less than p_i, the first of these numbers. This requirement is complicated by the interplay of i with itself. A simple example will help us to see how it behaves.

Let us take $n = 5$, so that we are considering the 5 prime numbers 2, 3, 5, 7, 11. If we choose $p_i = 3$, we have $i = 2$, and the group of numbers $p_i, \cdots, p_n$ consists of 3, 5, 7, 11. This group consists of $n - i + 1 = 5 - 2 + 1 = 4$ numbers. The number of these numbers is not less than the first of them (4 is not less than $p_i = 3$), so we have chosen too small a value for i. If we increase i by 1, to $i = 3$, we have $p_i = 5$, and there are only 3 numbers 5, 7, 11. Since 3 is less than 5, this is a suitable choice for i. Obviously any larger choice of i would also satisfy (2).

The result that we shall prove in this section is that *if i is chosen as small as possible, satisfying the condition* (2), *then we will have*

(4) $$p_1 \cdots p_i < p_{i+1} \cdots p_n.$$

This is true for $n = 5$, as can be verified by multiplication, $2 \cdot 3 \cdot 5 < 7 \cdot 11$. In order to see that it holds for further n, we must see how the optimal i changes with increasing n.

For $n = 5$ we had $i = 3$, $p_i = 5$. The group of numbers $p_i, \cdots, p_n$ contained 3 numbers, 5, 7, 11. If we change from $n = 5$ to $n = 6$ we bring in another prime number 13. However, we need not change i, since the group of numbers $p_i, \cdots, p_n$ can contain 4 numbers. If we change to $n = 7$, we bring in the prime number 17. We must now increase i, since otherwise the group $p_i, \cdots, p_n$ will contain 5 numbers and 5 would not be less than $p_i = 5$. Consequently we must choose $i = 4$ if $n = 7$. Now $i = 4$ implies $p_i = 7$, and this will permit a group of 6 numbers, 7, 11, 13, 17, 19, 23. That is, $i = 4$ will suffice for $n = 7$, 8, 9. For $n = 10$, the value of i must be increased to $i = 5$. Then p_i will be 11. Since this represents a jump of 4 over the previous $p_i = 7$, this value of i will suffice for 5 values of n, for $n = 10$, 11, 12, 13, 14. The value of p_i is shown in the following table by means of bold face type:

$n = 5$:   2, 3, **5**, 7, 11
$n = 6$:   2, 3, **5**, 7, 11, 13
$n = 7$:   2, 3, 5, **7**, 11, 13, 17
$n = 8$:   2, 3, 5, **7**, 11, 13, 17, 19
$n = 9$:   2, 3, 5, **7**, 11, 13, 17, 19, 23
$n = 10$:   2, 3, 5, 7, **11**, 13, 17, 19, 23, 29
$n = 11$:   2, 3, 5, 7, **11**, 13, 17, 19, 23, 29, 31
$n = 12$:   2, 3, 5, 7, **11**, 13, 17, 19, 23, 29, 31, 37
$n = 13$:   2, 3, 5, 7, **11**, 13, 17, 19, 23, 29, 31, 37, 41
$n = 14$:   2, 3, 5, 7, **11**, 13, 17, 19, 23, 29, 31, 37, 41, 43
$n = 15$:   2, 3, 5, 7, 11, **13**, 17, 19, 23, 29, 31, 37, 41, 43, 47

. . . .

Each time that i increases by 1, the prime number p_i increases by at least 2, and this allows i to remain the same for the next 3 values of n. If it happens that p_i increases by more than 2, the value of i remains constant for more values of n.

Now that we have seen how the optimal i changes with increasing n, we can discuss the validity of (4). We already know that (4) is true for $n = 5$,

$$(5) \qquad\qquad 2 \cdot 3 \cdot 5 < 7 \cdot 11.$$

Going to $n = 6$, i does not change, and (4) asserts that

$$(6) \qquad\qquad 2 \cdot 3 \cdot 5 < 7 \cdot 11 \cdot 13,$$

and this is clearly true, since the right side of (5) has been increased.

Going from $n = 6$ to $n = 7$ is not quite so simple because i changes its value here. The number 7 moves from the right side to the left, and the new prime 17 appears on the right. What we wish to prove is that

$$(7) \qquad\qquad 2 \cdot 3 \cdot 5 \cdot 7 < 11 \cdot 13 \cdot 17.$$

Without actually making the computation, we cannot obtain (7) directly from (6). We could multiply the left side of (6) by the factor 7 and the right side by the larger factor 17, but this would yield

$$2 \cdot 3 \cdot 5 \cdot 7 < 7 \cdot 11 \cdot 13 \cdot 17,$$

in which the unwanted factor 7 appears on the right.

If we start with (5), however, we can verify (7) without computation. Multiplying the left side of (5) by $7 \cdot 7$, the right by $13 \cdot 17$, we obtain

$$2 \cdot 3 \cdot 5 \cdot 7 \cdot 7 < 7 \cdot 11 \cdot 13 \cdot 17.$$

If a factor 7 is canceled out we have the inequality (7).

This argument carries us from $n = 6$ to $n = 7$ without any actual computations. It depends on the fact that we can go back *two* steps to $n = 5$. Exactly the same argument can be carried out for every similar step from one value of n to the next. It depends only on the fact that whenever the value of i increases, it then remains the same for at least *two* consecutive values of n (we have seen that it remains the same for at least *three* and often more values of n). Therefore the inequality (4) is true for all $n = 5, 6, \cdots$.

3. If we multiply both sides of (4) by $p_1 \cdots p_i$, we obtain

$$(p_1 \cdots p_i)^2 < p_1 \cdots p_n$$

or

(8) $$p_1 \cdots p_i < \sqrt{p_1 \cdots p_n}.$$

This, combined with (3), gives the desired result (1) for $n = 5, 6, \cdots$. That (1) is also true for $n = 4$ is easily verified by multiplying it out.

4. Bonse carried the discussion a little further. The inequality

(9) $$p_{n+1} < \sqrt[3]{p_1 \cdots p_n},$$

for $n \geq 5$, can be proved by the same methods. The decisive point in the proof is the fact that the values of i, discussed at the end of § 2, actually remain the same for 3 values of n, and not merely for 2 steps, as in the proof just completed.

28. An Improved Inequality

In Chapter 27 we mentioned that Bonse's proof of (8) actually gives the better inequality (9). As a matter of fact, the addition of one simple idea will allow us to prove even a little more in one direction, although, as we shall see, we will lose something in another direction.

The new idea is this: If M is a number of the form $6m - 1$ (a multiple of 6 decreased by 1), then in the decomposition of M into prime factors there must appear at least one prime which is also of the same form, $6x - 1$. To see that this is true, we notice that every number is of one of the forms, $6x$, $6x - 1$, $6x - 2$, $6x - 3$, $6x - 4$, or $6x - 5$. Now $6x$, $6x - 2$ and $6x - 4$ are all even numbers, so 2 is the only possible prime among them. Also, $6x - 3$ is divisible by 3, so 3 is the only prime of this form. All that remain are $6x - 1$ and $6x - 5$, so every prime number is either one of these two forms or is 2 or 3. However, 2 and 3 cannot divide M, which is of the form $6m - 1$. Furthermore, the product of two numbers $6y - 5$ and $6z - 5$ is

$$(6y-5)(6z-5) = 36yz - 30y - 30z + 25 = 6(6yz - 5y - 5z + 5) - 5$$

which is of the same form again, so, in order for M to have the form $6m - 1$, its decomposition must contain at least one prime factor of the form $6x - 1$.

In order to obtain our new inequality, improving over Chapter 27, we do not use *all* the prime numbers $p_1, p_2, p_3, p_4, \cdots = 2, 3, 5, 7, \cdots$ as we did before, instead, we take only $q_1, q_2, q_3, q_4, \cdots = 2, 3, 5, 11, \cdots$. We take $q_1 = 2$, $q_2 = 3$ and for the remaining q's, the prime numbers of the form $6x - 1$.

Now we form the expressions $M_1, M_2, \cdots, M_{q_i}$ just as before, but we use the q's instead of p's:

$$M_1 = q_1 q_2 \cdots q_{i-1} \cdot 1 - 1,$$
$$M_2 = q_1 q_2 \cdots q_{i-1} \cdot 2 - 1,$$
$$M_3 = q_1 q_2 \cdots q_{i-1} \cdot 3 - 1,$$
$$\cdots\cdots\cdots\cdots\cdots\cdots\cdots\cdots\cdots\cdots$$
$$M_{q_i} = q_1 q_2 \cdots q_{i-1} \cdot q_i - 1.$$

The statements (a), (b), (c) that we made concerning the original expressions are still true if we only read q in place of p.

The next step needs a little explanation. Just as before, if

(2*) $$n - i + 1 < q_i,$$

then there is some M, call it M_h, that is not divisible by any of the primes $q_1, \cdots, q_n$. If M_h is a prime itself, it is a prime of the form $6x - 1$ which comes after all the $q_1, \cdots, q_n$. If M_h is not a prime, by our first remark there is at least one prime of the form $6x - 1$ that divides it. Since this prime cannot be any of $q_1, \cdots, q_n$, it comes after q_n. In either case we find that there is a q of the form $6x - 1$ that divides M_h and comes after q_n. Since q_{n+1} is the next prime of the form $6x - 1$ that follows q_n, we have $q_{n+1} \leqq q$ and, since q divides M_h we have $q \leqq M_h < q_1 \cdots q_i$. Combining these results, we have

(3*) $$q_{n+1} < q_1 \cdots q_i$$

if (2*) holds.

Now, taking the smallest possible value for i still satisfying (2*), we make a table showing the value of q_i for each n, as we did in § 2. Since the table becomes rather large, we shall write down only certain parts of it:

$n=$ 6: 2, 3, **5**, 11, 17, 23

$n=$ 7: 2, 3, 5, **11**, 17, 23, 29

$n=$ 8: 2, 3, 5, **11**, 17, 23, 29, $\cdots$

$n=$ 9: 2, 3, 5, **11**, 17, 23, 29, $\cdots$

$n=$ 10: 2, 3, 5, **11**, 17, 23, 29, $\cdots$

$n=$ 11: 2, 3, 5, **11**, 17, 23, 29, $\cdots$

$n=$ 12: 2, 3, 5, **11**, 17, 23, 29, $\cdots$

$n=$ 13: 2, 3, 5, **11**, 17, 23, 29, $\cdots$

$n=$ 14: 2, 3, 5, 11, **17**, 23, 29, $\cdots$

. . . .

$n=$ 20: 2, 3, 5, 11, **17**, 23, 29, $\cdots$

$n=$ 21: 2, 3, 5, 11, 17, **23**, 29, $\cdots$

. . . .

$n=114$: 2, 3, 5, 11, 17, 23, 29, 41, 47, 53, 59, 71, 83, 89, **101**, 107, 113, $\cdots$

$n=115$: 2, 3, 5, 11, 17, 23, 29, 41, 47, 53, 59, 71, 83, 89, 101, **107**, 113, $\cdots$

. . . .

$n=121$: 2, 3, 5, 11, 17, 23, 29, 41, 47, 53, 59, 71, 83, 89, 101, **107**, 113, $\cdots$

$n=122$: 2, 3, 5, 11, 17, 23, 29, 41, 47, 53, 59, 71, 83, 89, 101, 107, **113**, $\cdots$

. . . .

The first thing to notice about this table is that whenever q_i increases, it increases by at least 6. This allows i to remain the same for at least the next 7 values of n. Because of the long pause that i makes, we will be able to prove

$$(4^*) \qquad\qquad (q_1 \cdots q_i)^6 < q_{i+1} \cdots q_n.$$

In fact, if this is true for part of the table up to any value of n, and if i does not change for the next value $n + 1$, then (4^*) still holds for $n + 1$, since only the right side is increased. If i does change, however, we can go back 7 steps to $n - 6$ without changing i and still have

$$(q_1 \cdots q_i)^6 < q_{i+1} \cdots q_{n-6}.$$

If we multiply the left side by q_i^7 and the right by

$$q_{n-5} q_{n-4} q_{n-3} q_{n-2} q_{n-1} q_n q_{n+1},$$

which is larger, we obtain

$$(q_1 \cdots q_i)^6 q_{i+1}^7 < q_{i+1} \cdots q_{n+1}.$$

Dividing both sides by q_{i+1} gives us our inequality for $n + 1$:

$$(q_1 \cdots q_{i+1})^6 < q_{i+2} \cdots q_{n+1}.$$

We have still not completely proved (4*). We have only seen that it is true for the value $n + 1$, provided either that it is true for n and i does not change, or that it is true for $n - 6$ and i does change. We shall see that it is true for $n = 114$ and 115. Since $n = 115$ is the first of at least 7 steps in which i does not change, (4*) will then be proved for all $n \geq 114$. It is easy to see that (4*) is true for $n = 114$ and 115. For $n = 115$ the left side is the 6-th power of a product of $i = 16$ factors; that is, the left side is the product of $6 \cdot 16 = 96$ factors. The right side is the product of $n - i = 115 - 16 = 99$ factors, and each factor on the right is larger than each one on the left. The proof for $n = 114$ is similar. This proof is very crude in that it does not make use of the fact that the numbers on the left are much smaller than those on the right. By making more careful computations we could probably reduce the size of n somewhat, but the computations would be quite lengthy. One fact is fairly obvious: (4*) is *not* true for $n = 6$.

If we now multiply both sides of (4*) by $q_1 \cdots q_i$ and take the 7-th root, we have

$$(8^*) \qquad q_1 \cdots q_i < \sqrt[7]{q_1 \cdots q_n}$$

if $n \geq 114$. Combining this with (3*), we have

$$(9^*) \qquad q_{n+1} < \sqrt[7]{q_1 \cdots q_n}, \quad n \geq 114.$$

This compares with Bonse's inequalities (1) and (9) of Chapter **27**, but it is proved for our special primes $q_1, q_2, \cdots$.

In order to get an inequality like (9*) for the set of *all* primes $p_1, p_2, \cdots$, we first notice that our proof of (9*) shows that the sequence of special primes q is not bounded. Therefore, if p_r is any prime, we can find a q_n such that $q_n \leq p_r < q_{n+1}$. Since p_{r+1} is the first prime of any sort that comes after p_r, we have $p_{r+1} \leq q_{n+1}$. Also $q_1 \cdots q_n \leq p_1 \cdots p_r$, since the right side includes all the primes $q_1, \cdots, q_n$, as well as some that are not of this form. Combining these facts with (9*), we have

$$(9^{**}) \qquad p_{r+1} < \sqrt[7]{p_1 \cdots p_r}.$$

We have proved this only for $p_r \geq q_n$, $n \geq 114$, that is, for $p_r \geq q_{114}$. This is our improvement of Bonse's inequality. We can say that (9**) is true if r is large enough. By checking through a list of prime numbers, we could find q_{114} and could then say just how large r must be. Even then (9**) would still probably be true for many smaller values of r, and these could be determined by actual computation. However, the main interest of (9**) is that it is true as soon as r passes a certain value and is always true from then on. The exact value of this r is of less interest.

In proving (9**) we have improved Bonse's inequality by replacing the square or cube root by the seventh root, which is considerably smaller. However, we have lost something in the process. Bonse's inequalities are true for all except the very smallest values of n, $n \geqq 4$ and $n \geqq 5$, while we have proved (9**) only for large values of r.

The reader should recognize that this discussion has required no mathematical knowledge other than the very simplest fundamentals, which were also used in Chapter 1. The proof depends entirely on pure reasoning. Because of this it shows clearly how ingenious and how difficult mathematics can be. In some cases it reaches its goal by combining and extending its numerous branches, but it also reveals its true spirit in examples such as this one, where the argument is developed with the aid of a very minimum of mathematical knowledge. If this last chapter seems to require a difficult chain of thought, if it shows how mathematics can build a real and meaningful structure on such a small foundation, then it probably exhibits most clearly the real motive of this book.

Notes and Remarks

CHAPTER 1

In the words of Euclid (*Elements* IX, 20), "Prime numbers are more than any assigned multitude of prime numbers."

For the proof concerning the gaps in the series of primes see Kronecker, *Vorl. über Zahlentheorie* I, Leipzig 1901, p. 68.

The existence of infinitely many primes in some other sequences, for example 1, 5, 9, 13, $\cdots$, $4n + 1$, $\cdots$ or 3, 7, 11, 15, $\cdots$, $4n - 1$, $\cdots$, both with common difference 4, can also be proved by elementary means. However, the sequence 2, 6, 10, $\cdots$, $4n + 2$, $\cdots$, with the same difference 4, contains only the single prime number 2 because all of its terms are even. In general, the theorem holds for series having any common difference if the first term is relatively prime to the difference. This was proved by means of higher mathematics in a famous and difficult paper by Dirichlet (1837), (*Abh. d. preuss. Ak. d. Wiss.* p. 45—81, or *Werke* I, p. 307—342),

CHAPTER 2

The representation of the networks as streetcar tracks may make it appear that the net of curves must lie in a plane. This is not necessary. All the results of this section are valid for networks of curves in space. This is in contrast to the subject of Chapter 10, which has a certain external similarity, but which does not carry over to space without change.

CHAPTER 3

§ 2. The discussion can be carried over directly from the circle to an ellipse. The role of the equilateral triangle is then taken over by the triangle for which the tangent at each vertex is parallel to the opposite side.

§ 3. The preliminary step was suggested by E. Steinitz.

CHAPTER 4

Anaxagoras—see Diels: *Die Fragmente der Vorsokratiker* I, no. 46, vol. 3, 2nd edition, p. 314_{16-19}.

Plato: *Laws*, VII, $819d_5$-$820c_9$.

The first proof is not to be found among the works of the Greek mathematicians, but it is the type of proof that they could have produced and they may have known it. The second proof is given by Euclid, *Elements* X, but an indication in Aristotle, *Analytica priora*, $41a_{26}$, $50a_{37}$, seems to show that it is older than Euclid.

CHAPTER 5 and 6

Schwarz, H. A.: *Ges. Abh.* II, p. 344-345; see also Steiner, J.: *Werke* II, p. 728-729; cf. also p. 95, no. 7 (= *Crelle*, 16, 1837, p. 88), where the assertion is given for both the plane and spherical triangle, and p. 238, 3.

Fejér's proof is not printed elsewhere. It has been reproduced with the kind permission of the author.

It is clearly necessary that the triangle have acute angles, since only in that case are all the altitudes inside the triangle. In an obtuse triangle the pedal triangle would not strictly be an inscribed triangle. In a right triangle, the pedal triangle reduces to a line.

In Schwarz's proof of the fact that the triangle is acute is used in assuming that the pedal triangle is inscribed in the original triangle. This is used in the reflected figure. In Fejér's proof the pedal triangle does not appear until the end. The acuteness of the triangle comes into play in keeping the angle $U'AU''$ less than two right angles so that the intersections M and N, of $U'U''$ with AC and AB, lie on these sides and not on their extensions. Furthermore because the triangle is acute, the foot E of the perpendicular is on the side BC and not on its extension.

An advantage of Fejér's proof is that it remains valid in non-Euclidean geometry. This is not true of Schwarz's proof, since it uses the fact that the sum of the angles of a triangles is 180°, as well as the Euclidean idea of parallel lines. In particular, since the sides of acute-angled spherical triangles are less than quadrants, Fejér's proof remains valid step for step in the case of a spherical triangle.

CHAPTER 8

Viggo Brun has used a more general combinatorial function relating to the number of combinations. See Netto, *Lehrbuch der Combinatorik*, 2nd edition, edited by V. Brun and Th. Skolem, Chap. 14, Leipzig, 1927. Brun has also discussed the particular function of this chapter in *L'Enseignement Mathématique*, vol. 29 (1930), p. 231-237.

CHAPTER 9

Bachet, in his edition of Diophantus' *Arithmetica*, has stated that a part of the *Arithmetica* implicitly includes the theorem on the representation of a number as a sum of four squares. Fermat gave a sketch of a proof in a letter, and Euler and Lagrange, *Works*, vol. 3, p. 189-201, 1869, proved the theorem in the manner indicated in the text.

Waring, *Meditationes Algebraicae*, 3rd edition, p. 349, Cambridge, 1782, conjectured that 9 cubes, 19 fourth powers, etc., will suffice. See also C. G. J. Jacobi, *Werke*, vol. 6, p. 322; Wieferich, *Math. Ann.*, vol. 66, p. 95-101, 106-108, 1909; Hilbert, *Math. Ann.*, vol. 67, p. 281-300, 1909, Hardy and Littlewood, *Math. Zeitschr.*, vol. 23, p. 1-37, 1925.

J. Liouville presented his proof at the Collège de France; it is printed by Lebesgue in the *Exercises d'analyse numérique*, p. 113-115, 1859.

The "etc." in Waring's conjecture may be interpreted to mean that any natural number n can be represented as a sum of

$$I = 2^k + q - 2$$

kth powers, where q is the greatest integer not surpassing $(3/2)^k$. So many kth powers are indeed needed for the number

$$n = 2^k q - 1$$

for which, since it is smaller than 3^k, only the summands 1^k and 2^k can be used. Now Dickson and Pillai proved in 1936 and the following years that for $k \geq 6$ the number I of kth powers indeed suffices for all n if (with Niven's later improvement)

$$\left(\frac{3}{2}\right)^k - q < 1 - \frac{q}{2^k}.$$

This condition holds for $2 \leq k \leq 200{,}000$ according to computations carried out by D. H. Lehmer, J. L. Selfridge and Rosemarie M. Stemmler. Moreover, we know also through the work of K. Mahler (1957) that at most finitely many k could violate this condition; no such k is known, however. For such exceptional k (if they exist at all) the number $g(k)$ of required kth powers has also been determined. For $k=2$, 3 the number I $(=4$, 9 respectively) is also good. But for $k=4$, 5 we know at present only

$$19 \leqq g(4) \leqq 35, \qquad 37 \leqq g(5) \leqq 54.$$

The work of Dickson and Pillai rests throughout on the results which Vinogradov obtained with his functiontheoretical methods that he developed from the methods of Hardy and Littlewood.

CHAPTER 10

§ 1. Gauss, *Werke*, vol. 8, p. 272, 282-286; see also Julius v. Sz. Nagy, *Math. Zeitschr.*, vol. 26, p. 579-592, especially p. 580-581, where the theorem is differently formulated and proved.

§ 4. It is not self-evident that a closed curve that is free of double points separates the plane into two regions. The theorem is not true for all surfaces. It fails, for example, on the torus. Therefore the theorem must be stated as being true *on the plane*, and it requires a proof. A proof was first given by C. Jordan, and the theorem now bears his name. Since Jordan's theorem does not hold for the torus (Fig. 117), our whole argument fails for this surface. In fact, a

Fig. 117   Fig. 118

curve can be drawn on the torus with two double points in the order 1212 (Fig. 118).

§ 5. Concerning alternating knots, see Tait, *Trans. Edinburgh Philos. Soc.*, vol. 28, p. 145, 1879.

The knot of Fig. 34 can be deformed, without tearing it, into the knot of Fig. 33, an alternating knot. However, it is not true that every knot can be deformed into an alternating knot. C. Bankwitz, *Math. Ann.*, vol. 103, p. 161, 1930, has given an example of a knot whose projection is never alternating.

Topics similar to those of §§2 and 10 are discussed by J. Petersen, *Acta. Math.*, vol. 15, p. 193-220, 1891.

CHAPTER 11

The theorem of the uniqueness of prime number factorization is not explicitly mentioned by Euclid. However he proves that if a product is divisible by a prime then at least one factor must be divisible by that prime (*Elements VII*, 24, 29), a theorem from which the uniqueness of prime factorization follows at once. For his proof Euclid makes use of the greatest common divisor, instead of the smallest common multiple.

A simple proof, using mathematical induction, has been found in the twentieth century (independently by Zermelo, Hasse, Lord Cherwell), which should be mentioned here.

The first few natural numbers certainly have a unique prime factorization. Indeed, leaving 1 aside as a unit, the numbers 2 and 3 are themselves prime numbers. Suppose there were numbers with two different prime factorizations, then there must exist a *smallest* among them, say

$$N = p_1 \cdot p_2 \cdots p_k = q_1 \cdot q_2 \cdots q_l$$

where the p and q are prime numbers, which in each product we assume ordered according to their magnitude

$$p_1 \leqq p_2 \leqq \cdots \leqq p_k \quad \text{and} \quad q_1 \leqq q_2 \leqq \cdots \leqq q_l.$$

All p must be different from all q, for if any $p_i = q_j$ then we could cancel these factors and would obtain N/p_i as a smaller number than N with two different prime factorizations. Without loss of generality we can assume $p_1 < q_1$. We then form

$$M = p_1 \cdot q_2 \cdots q_l < q_1 \cdot q_2 \cdots q_l = N.$$

This M is divisible by p_1, as N is. Therefore the difference $N^* = N - M$ is also divisible by p_1 and thus possesses a prime factorization which includes p_1. On the other hand we have

$$N^* = (q_1 - p_1)q_2 \cdots q_l$$

in which $q_1 - p_1$ is not divisible by p_1, nor is any of the primes $q_2, \cdots q_l$ equal to p_1. This means that N^* has also a prime factorization in which p_1 does not appear and thus possesses two different prime factorizations. But $N^* < N$ against our hypothesis that N is the smallest of such numbers. This contradiction disproves the existence of numbers possessing two different prime factorizations.

CHAPTER 12

The four-color problem was first mentioned by the mathematician and astronomer A. F. Moebius in a lecture in 1840. It became more widely known through a paper by A. Cayley in 1879, who emphasized its difficulty. In the same year A. B. Kempe published a paper containing an apparent proof of the theorem. However, his argument was shown to be faulty by P. J. Heawood in 1890. No proof of the four-color conjecture is known at present. However it is known that any map containing at most 37 colors can be colored by 4 colors.

A more comprehensive discussion of the four-color problem can be found in W. W. Rouse Ball, *Mathematical Recreations and Essays*, 11th edition, New York, 1939; and Philip Franklin, "The Four Color Problem," Galois Lectures, *Scripta Mathematica Library*, Nr. 5. (1941) pp. 49-85.

CHAPTER 13

§ 10. The properties of geometric figures which remain unchanged under distortions (but not tearing) form the subject matter of one branch of mathematics, *topology* or *analysis situs*. Our investigation of the regular polyhedrons is purely topological.

The idea of blowing the polyhedrons up into spheres represents a particular topological assumption. We will obtain entirely different results if we investigate polyhedrons that are topologically equivalent to a torus, and it will bring out the topological significance of §§ 7 to 10. It turns out that there are infinitely many "regular" maps (in the topological sense) on the surface of a torus, but that none can be realized as a regular polyhedron in the sense of metric geometry.

On the torus, Euler's formula becomes (see Chap. 12, § 5)

$$(3^*) \qquad\qquad v - e + f = 0.$$

Equations (4) and (5) obviously remain valid here, so we obtain, in place of (6),

$$(6^*) \qquad\qquad f(2\varphi + 2\varepsilon - \varphi\varepsilon) = 0,$$

and then, since $f \neq 0$,

$$(7^*) \qquad\qquad 2\varphi + 2\varepsilon - \varphi\varepsilon = 0,$$

or

$$(9^*) \qquad\qquad (\varphi - 2)(\varepsilon - 2) = 4.$$

Instead of the inequality (9) we now have an *equality*. The decomposition of 4 into two factors gives the possibilities $1 \cdot 4$, $2 \cdot 2$, and $4 \cdot 1$ for $(\varphi - 2) \cdot (\varepsilon - 2)$. For φ and ε, this gives the table:

φ	3	4	6
ε	6	4	3

These values satisfy (9*) and therefore (7*). But then (6*) is satisfied by *every* number f, since $f \cdot 0 = 0$ for all f. Consequently, in this case, f, v, and e cannot be found from the values of φ and ε.

Indeed, there are infinitely many values of f, v, and e for each pair φ, ε.

For the pair $\varphi = 4$, $\varepsilon = 4$ we can choose any number $a > 1$ and arrange a^2 squares in the form of a square (Fig. 119). This square can be rolled up into a cylinder (Fig. 120) and the cylinder can be bent into a torus (Fig. 121). Then the torus is covered with a regular map in which each country has $\varphi = 4$ boun-

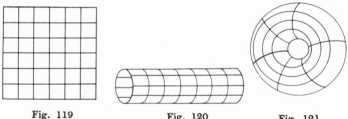

Fig. 119   Fig. 120   Fig. 121

daries, and $\varepsilon = 4$ countries touch at each vertex. Here we have $f = a^2$, and it is easy to see that we also have $v = a^2$, $e = 2a^2$. Since a is any number greater than 1, we have an infinite number of topologically regular maps on the torus. (It is more difficult to construct the maps for $\varphi = 6$, $\varepsilon = 3$. The case $f = 7$, $v = 14$, $e = 21$ appears among these maps. This is the map mentioned in Chap. 12, § 5, in connection with the coloring problem on the torus).

None of these polyhedrons can be realized as regular polyhedrons in the sense of metric geometriy. For if $\varphi = 4$, $\varepsilon = 4$, then the faces must be squares. Four squares around a point lie flat in a plane and do not form a three-dimensional vertex. No matter how many squares are put together, they will continue to lie in a plane and will not form the surface of a solid body. The same is true for $\varphi = 6$, $\varepsilon = 3$, since three regular hexagons around a point lie in a plane. Similarly, for $\varphi = 3$, $\varepsilon = 6$, six equilateral triangles around a point lie in a plane.

In summary, we notice *first* that the existence of just five topologically regular polyhedrons on a spherical surface is a topological property of the sphere. There are infinitely many topologically regular polyhedrons on a surface of the type of the torus. *Secondly*, topological regularity may not imply metric regularity. The regular polyhedrons of the type of the torus cannot be realized as regular polyhedrons in the metric sense. If the five metrically regular polyhedrons on the sphere can be realized, it will be because of special metric properties of the sphere.

CHAPTER 14

§ 7. Leaving aside the case $n = 4$ (which is treated in § 8) it suffices to prove the impossibility of $x^n + y^n = z^n$ in integers for prime numbers $n = p$ only. In the literature two cases are distinguished: case I, in which $x \cdot y \cdot z$ is not divisible by p, case II, in which one (and only one) of the numbers x, y, z is divisible by p. In case I Fermat's conjecture has been proved for all prime numbers $p < 253, 747, 889$ (by D. H. Lehmer, Emma Lehmer, and Rosser). In case II deeper theorems of number theory have to be used, and here Fermat's conjecture has been verified for all $p \leqq 4001$, with the use of the SWAC, an electronic digital computing machine at Los Angeles, by D. H. and Emma Lehmer and Vandiver.

CHAPTER 16

H. W. E. Jung, *Jourual f.d. reine u. angew. Mathem.*, vol. 123, p. 241-257, 1901, investigates the analogous problem in n dimensions. He finds

$$r = d\sqrt{\frac{n}{2(n+1)}}$$

as the radius of the spanning sphere for a finite set of points in n dimensions, with span d. The case $n = 2$ is the problem of the text.

CHAPTER 18

See Gabriel Koenigs, *Leçons de Cinématique*, Paris 1897, p. 273-285.

CHAPTER 19

§ 7. The prime numbers of the form $2^p - 1$, where p itself is a prime number, are called Mersenne primes (after Père Marin Mersenne, a correspondent of Fermat and Descartes). Those known at present belong to
$p = 2, 3, 5, 7, 13, 17, 19, 31, 61, 89, 107, 127, 521, 607, 1279, 2203, 2281$,
of which the last five were found in 1952−1953 by means of the digital computer SWAC in Los Angeles. Each Mersenne prime produces a perfect number.

CHAPTER 22

J. Steiner, *Werke*, vol. II, p. 193-195. The completion of the proof indicated at the end of the chapter has been given by C. Carathéodory and E. Study, *Math. Ann.*, vol. 68, p. 133-140. For a different, complete proof see Edler, *Göttinger Nachrichten*, 1882, p. 73.

A remark in Simplikios' commentary on Aristotle's *De Coelo*, *Berl. Ak. Ausg.*

VII, p. 412₁₃, states that Archimedes and Zenodorus had proved the theorem for both plane and solid figures. It also asserts that the theorem was known even before Aristotle's time.

CHAPTER 23

Gauss, *Disquisitiones Arithmeticae*, art. 312-318, discusses periodic decimal fractions, but he makes use of many results from number theory.

§ 4. The number $\varphi(n)$ depends upon n. It is frequently called Euler's function, and may be defined as the number of reduced proper fractions of denominator n.

§ 8. If p, q, r, $\cdots$ are the *different* prime numbers which divide n, then $\varphi(n)$ is given by the formula

$$\varphi(n) = n \cdot \frac{p-1}{p} \cdot \frac{q-1}{q} \cdot \frac{r-1}{r} \cdots.$$

The proof of this formula can be found in any textbook of the theory of numbers, for example, Hardy and Wright, *An Introduction to the Theory of Numbers*, Oxford 1938, p. 64, theorem 63.

CHAPTER 25

§ 7. The property of a curve of constant breadth asserted by theorem IV also holds for all convex curves, and it can be proved directly. However, the proof makes use of limiting processes and is best carried out in its analytical formulation.

The part of the theory of curves of constant breadth which can be handled by elementary methods is completely given in this chapter. There is a considerable body of literature concerning the more difficult problems of the theory. The reader who is familiar with integral and differential calculus may be referred to W. Blaschke, *Kreis and Kugel*, Leipzig, 1916.

CHAPTER 26

§ 1. Lorenzo Mascheroni (1750-1800) states, in the introduction to his book *La geometria del compasso* (Pavia, 1797), that he had studied the question of constructions with the compass alone, originally for practical reasons. Actual constructions made with a compass are usually more accurate than those made with a straightedge, a fact, as Mascheroni knew, that is used by astronomers when they wish to produce accurate scales on their instruments. His thorough study of constructions with the compass led him to discover that all the constructions of Euclidean geometry can be carried out with the compass alone.

Mascheroni dedicated his book to Napoleon, praising him as the liberator of Northern Italy. In return, Napoleon brought the book to the attention of French scholars in a conversation (December 10, 1797) with members of the French Academy. In the French translation (*Géométrie du Compas*, 2nd edition, Paris, 1828) this conversation is described as follows:

"Lagrange et Laplace faisaient partie de la réunion, et dans une conversation que Bonaparte eut avec ces illustres géomètres, et particulièrement avec Laplace il leur fit connaître la Géométrie du Compas, ouvrage alors tout nòuveau et inconnu en France, en leur donnant la solution de quelques-uns des problèmes qui se trouvent dans cette production originale. Après avoir écouté Bonaparte avec attention, Laplace, qui avait été son professeur de Mathématiques à l'école de Brienne, lui dit en présence de tous les savans réunis autour d'eux: 'Nous attendions tout de vous, général, excepté des leçons de Mathématiques'."

The only known work of the Danish mathematician Georg Mohr (1640-1697) is his *Euclides Danicus* (Amsterdam 1672), and this has only recently been rediscovered, (reprinted with German translation Copenhagen, 1928). The first part of this book is devoted to the problem of geometric constructions.

Jacob Steiner (1796-1863), Swiss by birth, was a professor at the University of Berlin. The problem mentioned in this chapter is taken up in his book "Die geometrischen Konstructionen, ausgeführt mittelst der geraden Linie und eines festen Kreises, als Lehrgegenstand auf höheren Unterrichts-Anstalten und zur praktischen Benutzung" (Berlin, 1833).

The problem as to whether the center of a circle can be found by means of a straightedge alone, and the method of solution, go back to David Hilbert. It was first published by one of his students, Detlef Cauer (*Math. Ann.*, vol. 73, p. 90-94, 1912, and vol. 74, p. 462-464, 1913).

§§ 6 and 7. We have considered only the case of two *non*-intersecting circles. It would be impossible to find a projection of the type we have used for two intersecting circles. In fact, such a projection would prove that the centers of the circles could not be found by means of constructions with the straightedge alone but, as was mentioned in § 1, such a construction can be found in this case.

The construction is really quite simple. In the first place, Fig. 122 shows how a diameter of a circle can be found if we are given two parallel chords AA' and CC'. Because of symmetry, the line DD' passes through the center of the circle. If a second pair of parallel chords is also known, then a second diameter can be drawn and the center is determined by the intersection of the two diameters. Now we must find a method of constructing two parallel chords. This is shown in Fig. 123. The line AA' is any chord of one of the intersecting circles. Starting at A, we draw the line ASB and the line $BS'C$, thus determining the point C.

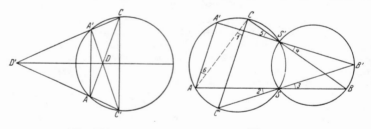

Fig. 122   Fig. 123

Similarly, starting at A' we draw $A'S'B'$ and then $B'SC'$. The points C and C' determine another chord. To prove that AA' and CC' are parallel, we need only show that the alternate interior angles 1 and 6 are equal. Going through the angles 1, 2, $\cdots$ 6 in turn, we see that each is equal to the next, either because they intercept the same arc or because they are vertical angles.

A construction to determine the centers of three non-intersecting circles is somewhat more complicated. We shall not reproduce it here, since it requires a more thorough knowledge of geometry. A construction may be found in Cauer's paper mentioned above.

§ 8. Other, shorter proofs of the theorem can be found, but they do not bring out the purely geometric aspects as well as the proof we have given. We chose this proof especially because it emphasizes the essential part of the theorem, the symmetry of the oblique cone.

CHAPTER 27

Bonse, *Archiv. d. Math. u. Phys.* (3), vol. 12, p. 292-295, 1907; cf. also R. Remak, *Archiv d. Math. u. Phys.* (3), vol. 15, p. 186-193, 1908.

Actually, much more than $p_{n+1} < 2p_n$ is now known. However, the best results are still not good enough to determine whether there is always a prime number between every pair of consecutive square numbers, for example, between 100 and 121, between 121 and 144, and so on.